1/15/-

In Quest of Man

In Quest of Man

A BIOLOGICAL APPROACH TO THE PROBLEM
OF MAN'S PLACE IN NATURE

by

PAUL ALSBERG

PERGAMON PRESS

Oxford · New York · Toronto
Sydney · Braunschweig

Pergamon Press Ltd., Headington Hill Hall, Oxford

Pergamon Press Inc., Maxwell House, Fairview Park, Elmsford, New York 10523

Pergamon of Canada Ltd., 207 Queen's Quay West, Toronto 1

Pergamon Press (Aust.) Pty. Ltd., 19a Boundary Street,
Rushcutters Bay, N.S.W. 2011, Australia

Vieweg & Sohn GmbH, Burgplatz 1, Braunschweig

First edition 1970

Library of Congress Catalog Card No. 70–112306

Printed in Great Britain by Cox & Wyman Ltd., London, Fakenham and Reading

Contents

List of Figures and Plates

Foreword

by Professor C. H. Waddington, c.b.e., m.a., sc.d., f.r.s.
The University of Edinburgh

MAN is a living creature, and therefore an object which comes under the biologist's scrutiny. Dr. Alsberg was a biologist, and very properly subtitled his book "A biological approach to the problem of Man's place in Nature". But he was a biologist with a broad enough point of view to realize not only the strength but also the limitations of our biological understanding. We find ourselves in fact in the paradoxical position that the more we understand about the fundamental nature of the biological world in general, the less, we realize, do we really comprehend the nature of Man, since it becomes ever more clear that there is an overwhelmingly important dividing line between man and the other animals.

Darwin's theory of evolution, elaborated and refined, but not altered in any of its essential implications, by the discoveries of modern genetics, provides the basic framework for our understanding of living things in general. Their nature is dependent on the fact that they have come into existence as the result of random changes in previously existing creatures and have had to pass the test of being able to make a living for themselves in a competitive world. We know in considerable detail how all this comes about, and have a technical vocabulary in which to express it: genes, the laws of inheritance, mutation, natural selection, and so on. But the more we know about this, the more obvious it becomes that the nature of man, although still involving all these elements, depends on something else besides. The fact that Man can fly does not depend, at least in any clear-cut or simple way, on a process of gene mutation and natural selection. Man has invented—or developed, or evolved, or let us settle for the neutral term "acquired"—a new method of evolution.

The essential nature of all that is most interesting about him comes into being by processes quite different from those which give rise to the characteristics of other animals. His eyes, muscles, digestive tract and genitalia are the outcome of evolutionary processes of the same kind as those which affect the Chimpanzee and the Gorilla; Man's painting, dancing, gastronomy and poetry are quite another kettle of fish, caught in a different stream, into which no other animal has yet been able to throw its nets.

As Alsberg spelt out in detail, Man's essential nature is determined by the fact that he takes part in cultural evolution as well as in biological evolution. Participating in this new method of formation—the fact that Man's clay is moulded by a potter who works in quite a different way—sets the human species quite distinctively apart from all the rest of the animal world. Man is biologically unique. He is, *essentially*, a product of culture; though of course built on a structure of rudiments produced by natural evolution. All other animals are essentially products of evolution, though some of them may be slightly modified by adumbrations of the cultural process.

In making so forcefully the point that Man's nature is moulded by forces other than those which shape the nature of other animals, Alsberg added his voice to those of a number of recent biologists, such as Julian Huxley, Mayr, Simpson, Dobhansky and others including myself. And perhaps I may be allowed to say that I do not altogether agree with, but was interested and stimulated by, the way he expressed the difference between Man and Animal. Alsberg put it in terms of the end that is achieved. In animal evolution, he said, the physical body becomes modified so as to be adapted to the circumstances it has to deal with—the principle is one of "body-compulsion"—whereas human cultural evolution consists in the development of "tools" (using the word in a broad sense) to which the demands made on the body by circumstances can be delegated. We do not need a heavy fist if we can press the switch which operates a pneumatic hammer, or to grow a lot of hair if we can set the thermostat on our air-conditioning plant. The principle is one of "body-liberation".

Before reading Alsberg's book, I had not thought of it in quite this way. I had been more concerned with contrasting the methods of operation of natural and cultural evolution. We know a lot about how genes are passed from one generation to the next, how they are changed by mutation and how this is affected by natural selection; much less about the equivalent cultural processes. Alsberg's emphasis on the kinds of result achieved by the two types of process raises many fascinating questions, which, I confess, I have not yet thought out at all clearly. What shall we do with our "liberated" bodies? Play with them? Which means starting up new lines of evolution in freely chosen directions. Or let them run down, to the minimum necessary to put together the apparatus and pull the switches?

Alsberg's book has two of the greatest merits that one can look for in a work of this kind. It makes, convincingly, a true and important point which many people have not yet realized, and it raises a lot of questions which no one can yet answer.

Preface

THANKS are expressed to F. Colin Sutton, who revised the whole manuscript. The task he has accomplished in improving the style and grammar of the text is gratefully acknowledged.

Part One

THE THEORY OF EVOLUTION
AND MAN

Introduction

THE standpoint taken in this book is that Man arose from low beginnings and in his evolution gradually proceeded to the present state of high cultural achievement; yet that nevertheless, his singular mode of evolution, as evidenced by the unique phenomenon of civilization, has set up an unbridgeable gulf between him and his animal ancestry. The author thus follows the Darwinian theory regarding the animal origin of Man, but challenges its implication that for his animal origin Man should be classed with the Animal. He insists, rather, that the phenomenon of civilization is so exclusively distinctive of Man and so totally alien to the Animal, that no matter from where Man has come, he to all intents and purposes constitutes a category of his own.

The case of Man is indeed most extraordinary. Quite probably he "came from the Ape", and yet palpably he is not an Ape but in Darwin's own words "the wonder and glory of the Universe". The postulation, therefore, that his simian origin relegates Man to the animal class may perhaps appeal strongly to our logical sense, yet must be equally at variance with the distinctive facts of civilization. No wonder that the inference was regarded in not a few quarters as denigrating the "dignity of Man", and aroused such a storm of bitter and violent criticism that the whole theory of evolution was liable to fall into disrepute.

To fit in the obvious fact of Man's uniqueness with the evidence of his animal origin is the main problem that Darwin left for his followers to resolve, and so long as it remains in our minds with no satisfactory solution in sight, the "Battle of Darwinism" is unlikely to come to an end; nor will it be possible for the theory of evolution itself to rise to the profound significance that it implies and to obtain the full recognition that it deserves. On the other hand, it is a problem of great intricacy and one that has puzzled, and often defeated, many an able mind.

If a plausible answer could be given to this intricate problem, one which would reconcile the established theory of the animal descent of Man with the distinctive facts of his cultural life, then the gain would be twofold in that not only the case of Darwinians would be promoted effectively, but at the same time a valuable contribution would be made to appease all those who still believe in the "high destiny and nature of Man".

Such being the issue, the present work ventures upon a fresh approach along *biological* lines to the perplexing problem of Man, whilst it abstains from any philosophical or metaphysical approach as beyond its scope. It is proposed:

(1) to investigate how the old Darwinian school came to be persuaded into the belief of the animal rank of Man, and how the opposing school of thought failed to expose this fallacy;

(2) to state the principle that started and governed the specific process of human cultural evolution and by its fundamental difference from the governing principle of animal evolution drew a radical line of distinction between Man and Animal; and

(3) to discuss, in the new light of human evolution, Man's place in Nature, his evolutionary career, and his natural destination.

By theorizing on a subject as speculative as the hypothetical beginnings of Man one naturally runs the risk of criticism and of opposition. Still, it would have been only half the battle if we evaded facing the enigma as to how Man sprang from the Ape without being an Ape, and the Ape remained an Ape without becoming Man. The risk may perhaps be somewhat less now, as with the human principle taken as a guide we may be able to answer the crucial question: where does the Animal end and Man begin? This should also provide some valuable clue to the tricky problem of identifying early human fossils from animal relics.

The present work goes back to previous writings by the author, particularly his book *Das Menschheitsraetsel* (*The Riddle of Man*) published in Germany in 1922 and, in a revised edition, in Austria in 1937. As no English edition has yet come out the author deemed it justifiable to rewrite his thesis in the English language, hoping thereby to make a practical biological contribution to the solution of the problem of the origin of the human race and, in his attempt to reconcile the facts of science with the dictates of our conscience, to meet the need of many a reflective mind.

CHAPTER 1

The Theory of Evolution

ACCORDING to the theory of evolution life developed naturally on our planet in the course of long geological periods, stretching perhaps a thousand million years, and gradually rose from primitive organic structures to higher differentiated organisms up to the highest living form, Man. When proposed tentatively by Lamarck, and even when most brilliantly set out by Darwin, the theory of evolution was still far from being widely accepted. But in the meantime so large a body of converging and conclusive evidence has been accumulated that nowadays we no longer reasonably doubt that there has in fact taken place a universal process of evolution comprising and characterizing all life on earth.

Were it not for the implication that Man himself was involved in this process, there would, I think, never have been such bitter controversy about "Darwinism". But strange to say, it so happens that in the case of Man the evidence of his animal origin is not at all as difficult to furnish as might be expected considering the strong opposition which the theory of evolution once met, even in scientific circles. Not only are the structural elements of the human physique and also the main phases of human prehistory better known than those of any other species, but the difference which separates Man genetically from his presumably nearest animal kinship, the Manlike Apes, is strikingly small, the structural and functional resemblances here being indeed so close that the transformation of the simian into the human form is easy to demonstrate, particularly as with the greater antiquity of the human fossils the Apelike features are correspondingly increasing in number and extent.

7

While the question of the immediate animal precursor is still under discussion, there is today general agreement on this significant point that Man would never have sprung from one of the living anthropoids, but rather that he is linked to them by the bond of common ancestry. In other words, it is assumed that Man and the present Manlike Apes came from the same, however remote, ancestral stock. From this evolutionary scheme of a very remote common ancestry some authors have derived a certain comfort, it being a "very different thing" that Man descended from some extinct anthropoid form rather than from one of the hideous living creatures. On principle, I would think, it makes hardly any great difference whether Man came from this or from that Ape-form, if it were an *Ape* to whom he owes his origin. A more comforting view would seem to me to lie in the suggestion that Man's "low birth" may not at all be inconsistent with an idealistic outlook if only one is prepared to interpret the upward movement of the animate world in the sense of a process working towards the creation of Man as its material and spiritual crowning. This point will be discussed at greater length in the last chapter of this book. From whichever aspect, whether materialistic or idealistic, we may feel inclined to look upon the universal process of evolution, it is one and the same problem: we must accept the implication that Man is equally involved in this process, and that his origin has to be sought within the animal series.

Where, then, is his place in Nature? Is he, as the offspring of an Ape, of necessity himself an Ape, just as a child is of the same kind as his parents; or has he by his evolution risen to a higher level of animal life and thus still remains essentially an Animal; or is there some distinctive peculiarity about him that separates him altogether from the animal class? If this be so, what then is he?

Most evolutionists are agreed that Man, since he is originally evolved from the Animal and developed on the same evolutionary principles as the Animal, cannot claim any place in Nature other than among the Animals; yet because of his unique cultural achievements he would deserve the highest rank in the animal hierarchy. The conception of the zoological classification of Man goes back to Linnaeus's *Systema naturae* (1735) in which the human species was united with the Monkeys, Apes, and other Animals in the Order of the *Primates* on the purely morphological grounds of anatomical resemblances—whereas the modern scheme of classification associates Man with the Apes and Monkeys on the basis of their genetic relationship, thus changing their formal co-ordination adopted in the system of Linnaeus into a downright subordination of Man to the Apes. Small wonder, therefore, that in books dealing with evolution we may not infrequently come across such phrases as "the Apes including Man", or even "Man and the other Apes", as if Man's genuine place in Nature were legitimately among the Apes.

It should, however, be noted that the incorporation of Man in the animal kingdom is by no means an inherent postulate of the theory of evolution as such, since this theory is primarily concerned with problems of evolution and descent rather than with problems of rank and qualities. The paramount question, therefore, as to what Man essentially is by his nature—whether he is still an Animal or whether he has in his evolution become something else and thus would represent a new category of living being on the earth—is not touched upon, still less answered, by that theory itself. It follows that the proclamation of Man as being essentially an Animal rests solely on a logical inference which, although it may be strictly drawn from the theory of evolution, nevertheless denotes a *decisive step beyond the actual scope of the theory.*

To be sure, if the step is rightly taken, there can be no escape from it. Neither are we allowed to ignore, or reject, its implications for the only reason that we dislike them; nor are we to blame the followers of Darwin for having derived from the theory any possible logical inference without stopping short at Man as an alleged "exceptional case". Nor must we forget that the theory of evolution had set up a totally new situation, such as must necessarily call for a thorough revision of all the old concepts that Man had formed of his origin and his nature.

Indeed, after the traditional belief in the separate creation of
Man had pitifully collapsed under the pressure of the new theory,
the same doubt was bound to fall also upon the traditional belief
in Man's separate place in Nature. While in former times it was
a matter of irrefutable dogma that the erect gait, the power of
speech, and the faculty of reason separated Man strictly from the
rest of creation, it is now as irrefutably recognized that those outstanding achievements of Man were but the natural outcome of
a long evolutionary process which started from most primitive
conditions and only very gradually led up to what is summarily
called "civilization". No organ and no function could be detected
in Man which in a lower degree of development was not also
existent in the Animal. Moreover, the route taken by evolution
from the animal to the human stage was easy enough to imagine
by drawing a continuous line leading from the imperfect posture
of the Ape, over the stooping carriage of *Neanderthal Man*, to
the perfect upright gait of modern Man; from the clumsy use of
stones and of other natural objects by the Apes to the crudely
fashioned flints of early Man and finally to the brilliant tools and
devices of present-day technology; from the simple vocal sounds
of Animals and the primitive languages of backward human races
to the highly developed languages of civilized Man; and lastly,
from the low level of animal intellect to the highest levels of the
human mind. With regard to the extraordinary mental capacities
of Man, modern students of psychology were able to show that
even his supreme faculty of forming abstract ideas was based on
such mental functions as were also possessed by the Animal. It
was on such lines of reasoning that one arrived at the conclusion
that Man's natural place was within the animal series. Human
evolution, one would argue, is but a continuation and gradation
of animal evolution; for it was by continuous and gradual steps,
the one almost imperceptibly merging into the other, that evolution proceeded from the lower animal stage to the higher human
stage. Man, therefore, is different from the Animal only in
degree, not in kind.

At first sight, looking at the inference without prejudice, it would seem that this kind of reasoning is irrefutably logical. Man, since he is of animal extraction and by continuous and gradational evolution has acquired no single new organ or function and no quality other than is possessed in a lower degree also by the Animal, has in essence remained what his parent was, an Animal. The gap between him and the lower creation, however great it be today, is one of quantity only, not of quality. The inference is obvious and its apparent cogency makes it intelligible why many biologists are no longer willing to listen to the protestations of those who for a "mere logical abstraction" would not forsake their old belief in the "higher nature of Man".

Still, the obviousness of the inference, impressive as it certainly is, must not be allowed to dissuade us from carefully checking its accuracy, particularly as the inference is still strongly challenged by an opposing school of thought. This is not the place to expatiate on the acrimonious, often tumultuous, controversy which has been staged against "Darwinism" for its revolting "degradation of Man", and which would even nowadays occasionally flare up. One thing is certain that over a scientific doubt there is no point in "coming to blows". There are two alternatives: either the inference is held to be correct, when it has to be accepted willingly or unwillingly; or the inference is held to be wrong, and then its fallacy has to be exposed.

CHAPTER 2

The Battle for the Soul of Man

IN THIS precarious situation in which the strong logical claim that Man belongs to the animal series is challenged by an equally strong counter-claim contending that Man should be placed in a category of his own, the onus must rest on the ability of the latter school of thought to disprove the claim of the former.

What, then, is their counter-argument?

In substance it is this, that Man's sublime spiritual life is so exalted and exclusive a feature of his being that it carries in itself clear and definite evidence of his uniqueness in Nature and his distinction from the Animal. As a good instance of this type of argument we may quote here a "confession" attributed to A. R. Wallace, who himself was (with Darwin) a first exponent of the theory of evolution:

> As you know, I hold that there was a subsequent act of creation when Man had emerged from his ape-like ancestry, so that a spirit or soul was given to him. Nothing in evolution can account for the Soul of Man; the difference between Man and the other animals is unbridgeable: mathematics alone are sufficient to prove the possession of a faculty non-existent in other creatures. Then you have the artistic faculty. It is quite clear to me that the soul was a separate creation.

His statement at least shows that one may be ready to accept evolution as a fact, and yet shrink from the idea of placing Man among the Animals. It also shows how, for lack of proper evidence, the obvious mental supremacy of Man is conveniently taken to serve as a self-evident criterion for his fundamental separation from the Animal.

Basically, such broad reference to the exalted spiritual life of Man is not so very far remote from the old doctrine in which "rational" Man was strictly dissociated from the "irrational" Animal on the ground that it was his distinctive mental faculty of Reason that was the true source of all his higher cultural achievements including technology, language, philosophy, ethics, aesthetics, and so on.

It is well in line with such a trend of argument that, for the sake of saving the human soul, one would not even hesitate to sacrifice the human body. Obviously there was no escape from the proven evolutionary fact that the modern human form had developed from ancient human forms of pronounced Ape-like features. So, with respect to the body, evolution could not be denied and hence was readily accepted; however, on the stringent stipulation that the theory of evolution would apply only to the human body which by all means may be thrown to the dogs, that is, the animal world, but that the theory would never apply to the human mind which from the start was emphatically different in kind from the animal mind and thereby bestowed upon Man a separate place in Creation.

This, surely, is an interpretation which must break down under the weight of evidence. If evolution is acknowledged at all as a historical process, it would be utterly inconsistent only to accept evidence attesting to the evolution of the body and to ignore other evidence attesting just as strongly to the evolution of the mind; the more so, as the elaborate development of the human brain as a bodily organ must have been closely and insolubly connected with the development of the mental functions.

Other authors, in trying to find a way out of the dilemma, showed themselves willing to grant the human mind its original development from the animal mind, but insisted that the human mind had in the course of its development reached so high a level of perfection as eventually to have become totally different from the animal mind. As an instance of this sort of argument we may quote a passage from H. Bergson's book *Creative Evolution* (Engl. transl. by A. Mitchel, 1912):

The human brain is made like every brain to set up motor mechanisms and to enable us to choose among them, instantly, the one we shall put in motion. But it differs from other brains in this, that the number of mechanisms it can set up, and consequently the choice it gives as to which among them shall be released, is unlimited. Now, from the limited to the unlimited there is all the difference between the closed and the open. It is not a difference of degree, but of kind.

Even if one agrees that the higher spiritual capacities of Man created with him a new style of life on earth profoundly different from the conditions of animal life and hence a most valuable criterion for placing Man in a new category of existence, it is still difficult to understand why this momentous shift from the animal style to the human style should have been a mere matter of advanced degree. It is because of this inconclusive kind of argument that some authors put forward the suggestion that it must have been a new principle with which Man burst into the world, a principle, that is, which brought about his new style of life and the distinction of being "Man". This was indeed a very sound and promising suggestion which reached to the core of the problem. However, the results were rather disappointing. Man's power of "imagination", his "loss of instinct", his ability of "making experience", and other such vague qualities were proposed to be the underlying principle.

Altogether it would appear that the propositions put forth by the opposing school in their effort to prove Man's fundamental distinction from the Animal must lack the strength of factual and deductive evidence on which the Darwinian school was able to build its case. Even if it be conceded that Man's higher mental achievements, such as mathematics and metaphysics (which are obviously out of the reach of the Animal), are ground enough for granting him separate rank, this concession would still meet with criticism that those achievements are also out of the reach of the primitive human races.

Today, then, the situation in the battle for the soul of Man has not changed much since the time when E. Haeckel challenged his opponents with the words: "Either we take Reason in its wider sense, as it exists in the higher Mammals as well as in the majority

THE BATTLE FOR THE SOUL OF MAN

of Men; or we conceive it in its narrower sense, and then we miss it in the majority of Men as well as in most Animals."

Assuredly, should there be any principle, quality, or faculty which is peculiar to Man and which is presented as the source and foundation of his uniqueness and of his separate place in Nature, then such must be expected to comprise the whole body of Mankind, not merely the later phases of human evolution. Therefore the bare reference to the mental supremacy of civilized Man is not sufficient to disprove the other claim that Man, as he came and developed from the Animal, represents but a higher developmental stage of the Animal—a failure that must render the position of the "anti-materialistic" school pitifully weak. Indeed, whenever they ventured upon a major attack against the Darwinian citadel, they usually came off second best, the stronger weapons of logic and fact being manifestly on the other side.

CHAPTER 3

The Fallacious Inference

It is axiomatic that in human life we find certain features which are totally lacking in animal life, and though they seem to fit in harmoniously with the evolutionary scheme of gradual development they still cannot be denied an existence and significance of their own. It is the *phenomenon of civilization* which is a fascinating novelty on earth and at the same time characterizes and integrates a feature of human life which is both lacking in and alien to animal life. If, for simplification's sake, we reduce the highly complex concept of civilization to its basic elements— technology, speech, and conceptual thought or reason—the non-existence of civilization within the animal series may be broadly set out in the following short formula:

> *No Animal makes tools, no Animal speaks*
> *in words, or thinks in concepts.*

Here is a clear line of demarcation separating Man from the Animal—a dividing line which is safely drawn and only varies with the cultural level of civilization. For if we go backward in human evolution, we will come to a stage when conceptual thought had not yet dawned upon Mankind, as is still the case with certain primitive races of our own times. Going further back in human history we may presumably come upon an evolutionary stage in which even articulate speech was not yet born. Yet, however far back we may trace human evolution we cannot imagine early Man without his artificial tools. As the art of tool-making is unknown to the Animal and Man has always concentrated

16

upon the use of artificial tools, it follows that human life has been distinct from animal life since its inception.

The question now arises as to how the Darwinian doctrine, in presenting Man as but a higher developmental grade of the Animal, is able to be reconciled to the phenomenon of civilization. That there has been going on in human evolution a process of civilization in which the Animal has no share is too self-evident a fact to be overlooked, and certainly has not been disregarded by the old Darwinian school.

As has already been mentioned, the gulf between Man and the Animal, and their divergence in the conduct of life, although being fully appreciated, was not taken in an absolute sense, since it was believed that the gulf was by no means unbridgeable. So continuous and gradual, one would argue, was the passage from the animal to the human stage (the one stage almost imperceptibly merging into the following one) that no fundamental discrimination was logically permissible between the two stages. Accordingly, with the one stage being different from the other in degree only, all the major advances that lifted human life above animal life could equally be in degree only.

This kind of reasoning is indeed impressive, yet does not answer the question how it is that with the emergence of the new human stage the original animal stage is abruptly overcome, and we are left wondering whether *continuity* and *gradual advance* in evolution do actually prove what they are meant to prove, namely, that the new stage is different in degree only from the old one, and consequently Man himself is only a higher developmental stage of the Animal.

However small the evolutionary steps may have been when the one merged fluidly into the other, they merely show that evolution has been proceeding *continuously*, and since continuity is in itself already logically implied in the concept of evolution, we learn from the very gradual advances that there *was* evolution, and not separate creation. On the other hand, the evidence of a continuous line of gradual transitions is of eminent consequence in so far as it may well serve as a clue to the tricky problem:

wherefrom Man has come and whereto he is going. If we follow
the line backward, we note that it ends in the animal series, and
this gives us every reason to believe that Man has come from
there. Indeed, if there were not conclusive evidence that the
human stage can readily be derived from the animal stage on the
line of an unbroken sequence of gradual modifications, the theory
of the animal origin of Man would be deprived of its very basis.

Yet the issue here is not whether there were continuous passages
from the one stage to the other, but whether, in this unbroken
process, human evolution actually proceeded along the old animal
line, that is to say, whether it was the Animal which continued
on in Man and merely lifted him upon a higher developmental
stage of *animal* existence, or whether human evolution took a
course of its own different from the animal line and separating
Man from the Animal. If, to answer this question, we are following the line upward, we infallibly come across the phenomenon of
civilization which is non-existent in the animal world, and for
this reason must be suggestive of a new course of evolution in
Man.

To any such suggestion, however, the objection has been raised
that the detailed structure of the human body would attest
irrefutably to evolution having proceeded "in degree only". Not
only has the human body preserved its fundamental identity with
that of the Animal, but all the manifold modifications to which
it was submitted in the course of evolution were apparently modifications of degree only, some organs having become larger and
stronger, others smaller and weaker. For instance, if compared
with the corresponding organs of the Manlike Apes, the human
jaws and teeth appear to have decreased in size and strength,
whilst the human brain has increased in volume and elaboration.
There can indeed be no doubt about the fact that, although most
of the structural modifications of the human body are throughout
specific of Man, there is no human organ which does not also
exist in the Animal and would differ more than "in degree" from
the corresponding animal organ.

The criticism here is that such a jump from the part to the

whole is very unsafe, and is indeed strongly negatived by the obvious fact that the picture of the human body, taken as a whole, does by no means reveal a "difference in degree only" from that of the Manlike Apes, but is from it most dissimilar. Therefore, the old argument advanced by Haeckel and others, that the "unlikeness of the outward appearance of the human body" was simply due to the "divergent growth of the separate parts", can no longer satisfy us, since it is exactly this "divergence of growth" that is in question here and has to be accounted for.

It would be unbiological thinking to assume that a living organism is the mere sum of its constituent parts, or that the individual organs would grow haphazardly and independently, the one becoming larger, the other smaller. Rather have we to think of an organism in terms of a biological unit in which the several parts are in their varying structure and growth harmoniously and congenially correlated, so as to build up an integrated *organic whole*; and when we see that Man's most distinctive physical features—his perfect hand, his supporting foot, his short jaws and small teeth, his large and elaborate brain—are closely connected with his use of artificial tools, his upright gait, his speech and powers of thought, we again here come across the phenomenon of civilization which must have a definite relation to the "unlikeness of the human body".

Our inquiry has shown that whenever the phenomenon of civilization comes in, a distinct line is drawn against the Animal. This becomes the more obvious as the gap between Man and the Animal widens with the advancement of civilization. The line becomes less distinct as with going backwards the gap narrows, until in the actual transitional stage between Animal and Man it may be totally overlooked and lead to wrong conclusions.

As such an instance of a very gradual and delusive transition we may conveniently consider the very spectacle of Primeval Man emerging from his anthropoid precursor. So far we know next to nothing about this event; yet it is generally assumed that the first phase of human evolution would have evolved from the Ape's simple use of stones and of other objects as tools and weapons. If, on this assumption, we follow a suggestion made by Lubbock (quoted by Darwin), namely, that Primeval Man, when he first used flint stones for any purpose, would accidentally have splintered them and then used the sharp fragments, and that from this step it would be but a small one to break the flints on purpose, and not a very wide one to fashion them rudely—such a step-by-step transition from Ape to Man would lend itself readily to an interpretation along lines something like these:

1. The Ape (Man's anthropoid precursor) used stones and other objects for any purpose.

   Man (Primeval Man) likewise used stones and other objects for any purpose.

   From a logical point of view these are identical instances, differing at most in degree only.

2. Flints being liable to break with hard use, Man accidentally splintered them, and then used the sharp fragments, a certain advance in degree only.

3. Man came to break the stones on purpose—a further advance in degree.

4. Man, in time, improved upon his technique by rudely fashioning the flints—a distinct advance, but again one in degree only.

Here, then, we have a sequence of gradual stages leading, by small advances "in degree only", from the simple use of stones by the Ape to the making of primitive tools by Man—an unbroken sequence which would appear to force upon us the irrefutable inference that with such gradual advances there can be no difference other than in degree between Man himself and the Ape. Hence the gap between them must be taken as definitely bridged.

However, there is the complication that at the one end of the sequence stands the Ape who does not fashion stones and has no share in civilization, while at the other end stands tool-making Man who, even with his elementary technique of tool-shaping,

initiates the art of technology as a true manifestation of civilization. It shows that the gap between them is still open and by no
means bridged by the Animal.

If this be so we must suspect that a *gross error* has crept into
the inference, in which case the doctrine that human evolution
is merely a continuum of animal evolution, and Man only a
higher developmental degree of the Animal, is exploded.

CHAPTER 4

An Unsolved Problem

GLANCING again over the "battleground of Darwinism" it is now with different eyes that we look at the conflict. Two groups of reflective minds are ardently engaged in a bitter controversy. Each group believes firmly in the apparent soundness of its case, but neither of them has so far been able to confute the other. Each group boasts of carrying the sharper weapons in its hands, but both are actually fighting with blunt swords. No wonder that the contest still persists.

The minds comprising one group stand on the firm ground of the theory of evolution, and from the fundamental identity of human and animal bodily organs and functions and the evidence of a continuous pasage leading gradually from the animal to the human stage they rightly deduce the animal origin of Man. Preoccupied, however, by the idea that human and animal evolution are governed by the same principles, they use the evidence, which served them so well in proving the animal origin of Man, as proof also of a mere difference in degree between animal and human evolution, and subsequently arrive at the logical argument that Man is but a higher evolutionary degree of the Animal. This is a mistake which not only bars them from ascertaining the true nature and rank of Man, but at the same time catches them in the meshes of a logical inference which is palpably inconsistent with the facts of cultural civilization.

Those of the opposing group, on the other hand, stand on the firm ground of their own cultural life, and from the evidence of the non-existence of civilization in animal life they rightly deduce the uniqueness of Man and his privilege of a separate place in

Nature. Preoccupied, however, by the idea that the whole process of civilization, distinctive and exclusive as it is of Man, is primarily due to Man's higher intellect, his Reason, they logically conceive this supreme mental faculty to be the distinguishing gift which, as such, makes Man "Man" and separates him fundamentally from the Animal. But they are able neither to set out the basic principle through which the human mind should have become radically different from the animal mind, nor to reconcile their own conception of Man with the theory of evolution and its claim that Reason too had gradually evolved from the animal mind and this at a time when Man had already long been in existence. Thus they are caught up in the meshes of a dogma which only applies to the upper level of humanity and bars the lower human stages from their legitimate place within the human family as not yet being in the possession of Reason.

The first group proceeds from the Animal and neglects to follow the human path of evolution upward to its present stage of civilization, thus overlooking, and consequently denying, the sharp borderline which, on the evidence of the facts of civilization, separates Man from the Animal. The second group proceeds from civilized Man and neglects to pursue the path of human evolution downward to its low beginnings, thus drawing the borderline on much too high a plane, and consequently dropping a large section of primitive Mankind, still void of Reason, into the animal kingdom.

Both groups, then, have failed in their effort to solve the problem of Man regarding his relation to the Animal. Their failure is hardly surprising to anyone who keeps his mind open to both the facts of evolution and the facts of civilization. Darwinians, it is true, claim to be in the stronger position, and, as Haeckel's challenge shows, may easily put their finger on the weak spot of their opponents. However, they themselves get into insuperable difficulties when trying to fit in their "zoological" conception of Man with the facts of civilization although, in itself, it should be simple for them to find for Man his proper place within the animal class, if actually he was an Animal. Darwin was the first to grapple with this problem:

> As far as differences in certain important points of structure are concerned, man may no doubt rightly claim the rank of a Sub-order; and this rank is too low, if we look chiefly to his mental faculties. Nevertheless, from a genealogical point of view it appears that this rank is too high, and that man ought to form merely a Family, or possibly even only a Sub-family. . . . To attach much weight to the few but strong differences is the most obvious and perhaps the safest course, though it appears more correct to pay great attention to the many small resemblances, as giving a true natural classification.

In other words, the "true natural classification" of Man is impracticable, because there are "certain important points", some "few but strong differences" which do not readily allow the genealogical principle to be applied to Man. Some authors, therefore, in their effort to save the "zoological" classification of Man, would not even hesitate to put Man in an Order of his own alongside the Apes and Monkeys. This inconsistent expedient, which in fact kills the genealogical principle without solving the problem, only shows again that this principle cannot legitimately be applied to Man without offending against his "dignity" or, for that matter, against the facts of civilization.

From the failure of the two contesting groups we may learn that there is no profitable approach to the human problem unless both the facts of evolution and the facts of civilization are given equal weight. Generally speaking, it is in two ways that human evolution may be investigated. The one is to follow its course backward to its starting-point by concentrating upon any such fact that *links* Man with his ancestry so as to find out the source from which he sprang—his *descent*. The other is to follow the course from its start upward to its present stage of civilization by concentrating upon any fact that *separates* Man from his ancestry so as to find out what essentially he is or has become: his *nature*. Both ways have a significance and value of their own, and we cannot dispense with either. The first one has been successfully taken up by the Darwinian school. It thus remains for us to follow up the course in the opposite direction. But in going upward from the animal to the human stage we should not choose too small an interval between the two stages so as to avoid falling into the error of taking the continuous and gradual passage from the one stage to the other as a mere matter of "degree". On the other hand, if we make the interval wide enough, so as to be safe from the danger of overlooking the distinctive facts of civilization, we should not run into the other error of taking the higher cultural stage of human evolution as representing the very start and the only criterion of Man himself.

Our task thus will be to try to grasp the specific evolutionary principle which must be supposed to lie at the root of the unique process of human evolution, and to prove, on the one hand, that this principle comprises the totality of the distinctively human features and activities, that is to say, that it holds not only for the advanced stages of human evolution but equally also for its lower stages, and hence must have already been in operation when Primeval Man bravely started upon his human career; on the other hand, that the Animal, since it has no share in the phenomenon of civilization, must therefore be void of the basic principle on which civilization rests. This principle, then, if there be such principle, must be as exclusively distinctive of human evolution as it is different from the principle lying at the root of animal evolution. Thus it would account for the distinction and "higher nature" of Man as well as for the spectacular and otherwise unintelligible act of evolution that caused Man to spring from animal stock and yet cease to be an Animal when he followed his individual course of evolution—an adventurous course which was to lead him further and further away from and higher and higher above the lowlands of his animal precursors up to the heights of modern culture and civilization.

It is significant for the uncertainties in the "Battle of Darwinism" that the whole precarious situation has since the beginning of this century not changed appreciably, in spite of all the brilliant scientific progress made in the fields of biology, psychology, palaeontology, and so on. Obviously the battle is now less ardently fought on the back of science, although the actual mechanism of evolution is not yet properly recognized; but it is for the sake of the right *Weltanschauung* that it continues, and here the present book might well help to bring the two contesting schools of thought more closely together.

Part Two

THE PRINCIPLE OF HUMAN
EVOLUTION

CHAPTER 5

The Principle of Human Evolution

WHENEVER we turn our eyes and our minds to the multifarious world of the Animals, again and again we are deeply impressed by the spectacle of a countless variety of forms, structures, and contrivances, all of which make the Animals appear so perfectly fitted to their habitats and environments. It was mainly due to the theory of evolution that we became aware of the profound significance inherent in all those innumerable structural devices which combine to help the Animals in their hard struggle for existence by adapting them specifically to the various conditions of their surroundings.

Not only has the Bird wings to fly, the Fish fins to swim, and the Mammal legs to run; but we also find some Birds, like the Penguin, adjusted to aquatic life, and others, like the Ostrich, adjusted to terrestrial life. Some Fish, like the Flying Fish, are adjusted to flight through the air, and others, like the Mudskipper, are adjusted to living on the ground. Among the Mammals, some species, like the Whale, are adapted to the seas, and others, like the Bat, are adapted to the air.

In addition, we find contrivances which adjust the Animal to climatic conditions, such as a fur-coat, a feather-coat, or a blubber-layer; or devices which serve other particular adaptive ends, such as the burrowing feet of the Mole, the gnawing teeth of the Beaver, the spinning gland of the Spider, and the cheek-pouches of the Hamster. Equally impressive are the manifold structures used for attack, such as those of the carnivorous Animals equipped with formidable weapons such as the powerful teeth and claws of the Tiger, the hooked beak and sharp talons of the Hawk, and

29

the saw-teeth of the Shark. On the other hand, we find equally efficient means of defence, as for example the horns of the Bull, the tusks of the Boar, and the sting of the Bee. Again there are protective arrangements such as the shells of the Mollusc, the armour of the Tortoise, and the quills of the Porcupine; or swift feet to escape pursuit; or means of intimidation such as hissing, or snarling; or devices of deception such as imitative coloration and mimicry; or screening devices such as the sepia of the Cuttle Fish. Or, if the individual Animal is not itself sufficiently protected by its bodily make-up, the continuance of the species may be safeguarded by other adaptive means, such as prolific propagation.

There is indeed hardly a feature or design to be found in the Animal that is not in some way, directly or indirectly, related to the task of adaptation to environment and survival. The perfect adaptive outfit of the Animal is common knowledge, yet had to be stressed here for its close relation to the process of evolution. It was through the sifting and perpetuating agency of natural selection in the struggle for existence that all those manifold adaptive contrivances have developed to such perfection. Since survival of the individual, and indirectly of the species, is preeminently dependent upon their adaptability to environment, and since throughout the whole animal realm we find a common tendency towards making any possible use, for the same ends of adaptation, of any means and potentialities offered by the body, this universal and uniform tendency gives us a definite clue as to the specific scheme to which Animals are subjected in the course of their evolution.

We may safely say that animal evolution appears to be directed by a supreme principle which, in response to Nature's imperative demand of perfect adaptation to environment, keeps invariably to the means provided by the *body* when it seizes upon any structure or variation suitable for this purpose. In the case of the Animal, therefore, it is the *body* upon which evolution turns for the compulsory task of adaptation—with the ultimate result of that bewildering multitude of divergent adaptive forms. Hence, the principle which lies at the root of the Animal's evolutionary scheme presents itself as a *principle of body-adaptation*, that is, adaptation to environment by means of the body, and since this form of adaptation is the Animal's only available means of adaptation, the underlying compulsory principle may more appropriately be called *the principle of body-compulsion*.

We now turn to Man. How different is his picture from that of the Animal. It has been said that Man is "as naked as a Frog". This is in itself no good comparison, as the Frog is by no means "naked", but is in its own way perfectly well adapted to its environment. Man, however, is in every sense "naked". He can boast neither of powerful teeth nor of claws, nor of any other bodily weapon that would enable him to ward off dangerous enemies. He has neither feet swift enough to save his life by escape nor keen senses to discover danger from afar. Neither has he a thick hairy coat to keep off severe winter cold; nor indeed any such protective equipment that would enable him to cope successfully with the grim exigencies of life. Thus, completely lacking in all those adaptational devices upon which the Animal can confidently rely in its struggle for existence, Man appears in his evolution to be a singular exception to the universal scheme of body-compulsion.

His "nakedness", indeed, gives the impression of a most *unnatural* condition, and we cannot imagine any Animal being so utterly void of vital adaptive equipment. Forsaken in this way by Nature it would have no chance of survival. This reflection leads us to believe that the anthropoid precursor of Man, since he still was a true Animal, cannot possibly have been so acutely destitute of adaptive equipment, otherwise he would not have been able to maintain his own life, still less to give birth to the human line. But if he, like all Animals, was physically well adapted to his surroundings, then the present unnatural state of Man admits of only one conclusion: that those vital adaptations, still possessed by his animal precursor, must have been lost to him in the course of his evolution.

If this be so, the problem arises as to how to fit in his total loss of natural protective equipment with the exigencies of the

universal struggle for existence. It has been suggested that
Primeval Man may, for a while, have been exempted from that
fierce struggle if by chance he lived in a region free from
dangerous beasts, and here enjoyed a life of paradisial happiness

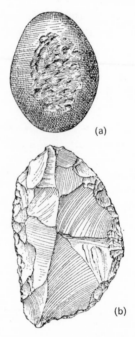

(a)

(b)

Fig. 1. (a) Small anvil or hammerstone, (b) oval flake-tool (flint). (By permission
of the Trustees of the British Museum (Natural History).)

with no need of defensive weapons; and that in consequence of
this peaceful existence, his original means of protection would
have gradually disintegrated, whilst his mental abilities would
have correspondingly increased. When in the course of time the
situation grew worse, and hardship and danger had again to be
faced, his intellect had meanwhile developed to such a degree that
it enabled him to make up for his physical deficiencies by *the
invention of artificial tools.*

This is a convenient, though unconvincing, theory. If we take conditions on the earth as they really are and presumably always have been—conditions of deadly competition—we cannot escape from the conclusion that Man, like the Animal, in every phase of his evolution must have been perfectly adapted to his surroundings and well equipped for the needs of existence. He never would have been able to survive and to secure the continuance

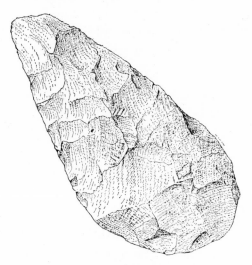

Fig. 2. Lava hand-axe. (By permission of the Trustees of the British Museum (Natural History).)

of his species, had he not been in possession of powerful protective weapons. Now, if those indispensable weapons are no longer found *within* his physical organization, they must be sought for from *without*. In point of fact, as far as we can trace back human history we find Man in the possession of *artificial tools*. The mental association between Man and his tools has become so firmly established that it is impossible to conceive Man without the possession of artificial tools, just as flints, however crude their flaking, are indicative of the presence of Man who fashioned them. It means that Man's strength and safety from the time of his emergence from Apehood have been resting upon his use of artificial tools.

If we follow this thought through to its logical implications, the problem of the "nakedness" of Man will soon be resolved. Primeval Man inherited from his animal predecessor, and hence originally possessed, a perfectly protective bodily outfit such as would have guaranteed his survival. However, when he took to

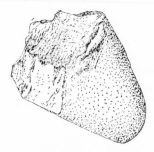

FIG. 3. Pebble-tool. (By permission of the Trustees of the British Museum (Natural History).)

using artificial tools for defence and other purposes instead of resorting to his natural protective means, he was no longer in urgent need of, and therefore neglected, his original adaptive equipment, which thus became liable to subsequent deterioration.

According to our theory, then, it was the adoption and perpetuation by Primeval Man of the practice of the use of tools which was primarily responsible for the decline of his original adaptive outfit. In this perspective of primary acquirement of artificial equipment and secondary loss of bodily equipment it would appear that Man has always been in the possession of effective means to hold his own in the struggle for existence.

Our theory provides a simple and yet adequate explanation of the strange contrast in the appearance of the well-equipped animal body and that of the non-equipped human body, in that their divergent development is here being taken as the direct result of two contrasting evolutionary schemes—that of the

Animal growing its adaptive means within the body, and Man taking them from without. While the animal scheme makes the body extremely fit to meet the exigencies of life, the human scheme leads the body into a state of extreme unfitness.

In the case of Man evolution appears to have taken a new direction in which adaptation to environment was no longer entrusted to the body but was implemented by *artificial tools*. In short, *bodily adaptation* switched over in the case of Man to *extra-bodily* adaptation. Consequently, with the duties of adaptation being transferred from the body to the artificial tool, human evolution took a course in which all effort that hitherto (in the animal ancestry) served the task of developing the highest possible grade of body-adaptation was henceforth (with Man) devoted to the highest possible development of the artificial tool. The final result on the one hand was the establishment of a gigantic realm of highly finished artificial tools and contrivances towering *around* Man, while on the other hand there was a singular deterioration of the body itself, which made his maintenance of life totally dependent upon the use of artificial tools.

That the artificial tools have their own existence *outside* the body is self-evident, as is also the fact that they are used *instead of the body*. Strangely enough, the whole issue has been obscured and indeed hopelessly confused by a theory, still widely maintained, that the tools would simply *enlarge* the capacities of the bodily organs. Thus, the hammer is said to be a copy of the clenched fist, and is intended to extend and reinforce the hand; the telescope, modelled after the eyes, to extend and reinforce the eyes, and so on. While in this way the artificial tools, though they are clearly and essentially of an *extra-bodily* order, are being happily transformed into "organ-projections", as *quasi-part* of our body, the situation becomes even more perplexed by the fiction that the (essentially bodily) organs of the Animal, such as the teeth and claws of the Tiger, represent the "natural *tools*" of the Animal.

Apparently an effort has here been made to equate the artificial tools of Man with the bodily organs of the Animal—an effort which may have been inspired by the wishful thought of bridging the gulf between tool-making Man and non-tool-making Animal, yet at bottom means nothing but forcing two totally different orders into one ill-contrived category.

The "organ-projection" theory has found so much favour because the organs, particularly hand and arm, are most often actively brought into play when using a tool—as, for example, in the act of swinging a hammer. Still more, the performance of the hammer is practically like that of an extended and reinforced hand. Yet, undeniably, the fact is that it is the hammer that drives the nail into the wall, not the hand; and that the hammer is even made for the special purpose of doing the work *instead of the hand*. Still the hammer cannot work unless it is put into operation, and it is only for this task of *operating* the hammer that the hand is here engaged. As to the actual work done and intended to be done by the hammer, the hand is, in fact, *eliminated*.

This seemingly paradoxical situation can easily be understood when one considers the case of the hammer being worked no longer by the hand but by some other means. Indeed, as our own physical exertion is required only for the purpose of *operating* the tool, the same purpose might as well be achieved by no organ at all, but by mechanical power. In the case of a hammer being worked by steam or electricity, nobody would hold the idea that our bodily capacities are hereby enlarged by the hammer. It would be equally senseless to suggest that our eyes are extended into heavenly space and that their vision is multiplied a thousand-fold in case the telescope is operated not by our eyes but by a photographic plate.

Why should Man altogether bother himself with artificial tools instead of using his own bodily powers, when he first has to make and then to operate them?

The reason is that the efficiency of artificial tools is in many respects infinitely superior to the abilities of the organs. Being of an extra-bodily order with an existence of their own, the tools can be used by anyone, in any number, in any combination and in any strength, as compared with the body-organs which are naturally confined to the one individual in which they exist and are limited to the extent of their number and strength.

Their other great advantage is that they can be made of any kind of material, and shaped in any form and size. This enormous variability must give them unlimited scope in their potentialities of composition, construction, and perfectibility, together with a high degree of precision, as compared with the body-organs which by their very nature are restricted within the narrow limits of their size, shape, and development. By means of artificial tools Man is indeed able to meet any possible requirement demanded by the occasion, and to cope with the varied and varying conditions of his surroundings in such a way and to such an extent that he would never have been able to do with his bodily resources, however highly developed. There is then every reason to believe that it was for their wider range and greater efficiency that Man has devoted himself to the contrivance and use of artificial tools. Thus he adhered to them and so made tool-use a permanent practice and one which was instinctively passed on to and further developed by each following generation. Tool use thus became the dynamic principle of human evolution.

It is not only size, shape, and construction that determine the specific performance of tools. They depend equally upon the raw material from which they are made. A tool of iron has an effect greatly differing from one made of wood, glass, or rubber, and it is for this reason that the material is specifically selected with a view to the work wanted from the tool. But whatever material is used for the fashioning of the tool—be it solid, fluid, or gas; or even if the material is derived from the body itself, such as bone, hair, or teeth, and which will determine the specific performance of the tool—it will never qualify the tool as such in its essence as an extra-bodily or artificial means of body elimination with an existence of its own. Thus in the case of the Bear his fur-coat is his self-grown means of bodily adaptation, whilst for Man who kills the Animal and hangs its skin over his shoulders to protect him from winter cold the same fur-coat is an extra-bodily means of adaptation, an artificial appliance, or "tool", that might be used by everyone for the same or other purpose.

The use of artificial tools does not simply mean replacing one means of adaptation, the body-organ, by another more efficient one, the extra-bodily tool. What it really means is that in human evolution *the task of adaptation to environment has switched over the body to the tool.* Consequently not only was the human body saved from those adaptational specializations that are apt to tie the Animal down to a specific environment and a fixed mode of life, but the higher capacities of the artificial tools, as compared with the narrowly limited capacities of the body-organs, have opened the way to achievements which are totally out of the reach of the Animal and allow Man to make his own choice of environment and of conduct, an altogether revolutionary change of behavioural style which, in the end, conferred upon Man lordship on earth.

From an evolutionary point of view it would appear that by renouncing his own bodily resources and choosing instead his extra-bodily scheme of adaptation *Man shook off the yoke of compulsory bodily adaptation* and thereby *freed* himself by means of his superior artificial tools from the natural restrictions and limitations inherent in the animal scheme of body-adaptation.

There is, then, a profound contrast between the schemes of animal and human evolution, which reveals itself distinctly in this short formula:

> The principle of animal evolution is that of compulsory adaptation to environment by means of the body: *the principle of body-compulsion.* The principle of human evolution is that of freeing Man from the compulsion of body-adaptation by means of artificial tools; *the principle of body-liberation.*

Our inquiry started from the strange phenomenon of a wide disparity existing between the perfect adaptive outfit of the animal body and the "nakedness" of the human body: and it concluded with the thesis that it was the elimination of the human body by means of artificial tools from the duty of adaptation that subsequently led to the loss of original adaptational equipment. However, as the making and the operation of tools is greatly dependent upon body-function, particularly in the early human stages when tools were still crude and clumsy to manage, we should expect that such new and constant demands made upon the body-organs must have had a markedly stimulating effect upon their development. Therefore, if the principle of body-liberation is in fact the dynamic principle of human evolution, we should find, apart from those regressive features due to

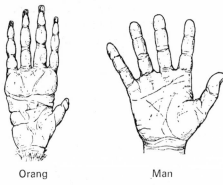

Orang   Man

FIG. 4. Hand of Orang and Man.

(From Mankind in the Making *by William Howells, drawings by Janis Cirulis. Copyright © 1959, 1967 by William Howells. Reprinted by permission of William Howells, Doubleday & Co., Inc., and Martin Secker & Warburg, Ltd.)*

body-elimination, definite signs of *progressive* development in all those organs which were pre-eminently engaged in the employment of tools.

Evidence of progressive development is indeed easy to furnish. Take, for example, the human hand, the organ foremost involved in the manufacture and employment of tools. Its extraordinary dexterity and versatility is so outstanding a feature of Man that his hand has rightly been called the "organ of organs". Still, its much-praised perfection must not be conceived in the sense that this was an original condition. Rather are we to believe, in accordance with the theory of evolution, that the human hand is derived from the climbing hand of the Ape; and although we may not choose as its prototype the excessively long and curved hands of the Orang or the Chimpanzee, nor even the less specialized hand of the Gorilla, the present shape of the human hand reveals by its greater width, its relative shortness, and, above all, the singularly well-developed thumb the unmistakable signs of a progressive development in specific response to tool-use. Yet, significantly enough, it is merely with relation to tool-use that the high "perfection" of the human hand holds good. With regard to

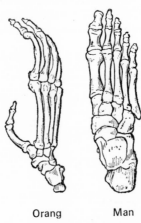

Orang   Man

Fig. 5. Foot of Orang and Man.

climbing the human hand is anything but perfect and cannot in the least compare with the Ape's hand, which is "perfect" too, though merely with respect to climbing. It would appear therefore that the present "perfection" of the human hand has been bought at the price of loss of its former climbing ability; and since climbing is a form of body-adaptation, the present state of the human hand suggests that this organ when developing in the direction of tool-use had by its corresponding loss of climbing fitness to take its full share in the general disintegration of the original adaptive outfit of the body.

This is equally true of the human foot. Here again we find progressive development in response to tool-use at the expense of former adaptive outfit. Derived from the supple prehensile foot of the Ape, the human foot developed into a rigid supporting organ which, with its shorter toes and its no longer opposable great toe, has lost its original climbing ability and has become instead eminently suited for erect walking, thus allowing free use of hands and arms in the employment of tools.

A further instance of progressive growth is the enormous expansion and elaboration of the human brain. As will be discussed later, its development from the smaller and more primitive simian brain must again be supposed to have been greatly influenced by the use of artificial tools; and here also it would seem progress was bought at a price: the decline of instinct as an element of former adaptive outfit.

The characteristic combination of both retrogressive and progressive features in close relation to the use of artificial tools must bring powerful evidence to bear on the principle of body-liberation reigning supreme within the process of human evolution. It is indeed only because this principle has impressed its distinctive stamp so deeply upon the architecture of the human body that the picture of Man differs widely and strikingly from that of the Animal. In the first part of this book we came to suspect the "unlikeness of the outward appearance of the human body" to be due to a new course of evolution. Now that we have grasped the principle that lies at the root of the process of human evolution, we can easily account for that "divergent growth". Evolution, hitherto tending to adapt the body to the requirements of environment, is now setting out to adapt the body to the requirements of tool-use. This change, of course, imposed upon the body a structure as divergent from the animal body as it is specific of Man.

Still, it is only the effect, however significant, of tool-use upon the *physical* development of Man that we can deduce from the specific structure of the human body. Our next task, therefore, must be to probe the human principle in its immediate fields of expression—that is, in its variety of tools and its various ways of body-liberation—so as to add direct to indirect evidence. Accordingly we must now turn from the specific form of the human *body* as the true reflection of the human principle to the specific form of human *life*, the life of civilization, as being the palpable *manifestation* of this principle.

Only if there be proof that the whole complex framework of civilized life with its outstanding achievements in both the technical and the mental spheres is in fact resting upon the properties and activities of the same principle of body-liberation, and that hence this principle is as distinctive of human evolution as it is alien to animal evolution, shall we have full certainty that the human process is actually and exclusively conditioned by the operation and development of this principle—which then will have proved itself to be the true principle of human evolution.

CHAPTER 6

Technology

IF THE concept "Technology" is taken in the sense of referring not only to the scientific basis but also to the practical application of the technical arts, one will agree that, so far as modern Technology is concerned, the human principle of body-liberation has already stood its test by what has been said in the previous chapter. Tools, we saw, are essentially extra-organismal, artificial means made and used for the ends of body-liberation. On this definition there is indeed hardly a technical or industrial product that would not serve, in some way or other, the same purpose and thus testify to the supremacy of the basic human principle.

A cursory glance over the conventional technical appliances in practical use everywhere among civilized Men suffices to bear out the dominating tendency in modern cultural life towards the advancement of body-liberation. We live in solid houses which are designed to protect us from the inclemencies of weather, and which are equipped with a variety of devices, such as fire-places or radiators to overcome the winter cold; lamps or candles to illuminate darkness; fans or ventilators to improve air-conditioning; and so on. We sit in chairs and sleep in beds; we wear clothes, shoes, hats, gloves; we carry a stick or an umbrella. We shoot game with a gun, and catch fish with net or hook; we chop or mince and cook meat, and eat with spoon, knife, and fork.

What is all this for?

Some would answer "because it is comfortable or fashionable or good manners". A cynic would perhaps insist that he is properly used to sit at table, to have his meals served on dishes and plates, and to eat his food with knife, fork and spoon because he is not a "pig"; and he indeed hits the nail right on the head in that his answer makes it clear that Man, unlike the Animal, tends taking to artificial tools and devices as an expression of his "true nature", that is, in deference to his principle of evolution.

How far Man has succeeded in his endeavour to make himself free from body-compulsion is best illustrated by some instances of peak-achievement of modern Technology. By railway, car, and boat he travels all over the globe; by aeroplane he moves through the air; by telephone he talks over thousands of miles; by radio he listens to a concert performed in a foreign country, and by television he even sees the orchestra itself. With the telescope he penetrates the immensity of the universe, and with the microscope he investigates the structure of cells and atoms. All this he achieves without growing, as would an Animal, such organs that would allow him to carry out those feats. It clearly attests to one and the same fact that Man tends to release his body from the environmental adaptation and insists upon meeting any condition in the world by means of artificial devices, and it is his acquirement of technological skills which places those efficient devices of artificial adaptation at his disposal.

The suggestion has been put forward that Man owes his singular independence of environment mainly to his greater *mental* powers, and that these enable him to free his body from those structural specializations that inevitably confine the Animal to a particular environment and to a correspondingly fixed mode of life. There is little doubt that the human intellect has steadily developed during the whole period of human evolution, and with its higher development has played an increasingly prominent part in the development of Technology; and that with science striving to set Technology upon a rational basis it is the power of abstract thought that is responsible for the growth of modern technical progress.

However, if we go backward in geological time to that critical point when Man abandoned the animal scheme of bodily adaptation by turning to artificial adaptation, and so saved his body from those restricting specializations—that is, when he actually started upon his career as "Man"—his mental powers would not yet have developed to any great extent. Moreover, Man is by no means completely free from such specializations. On the contrary, the human body is in many respects as highly specialized as that of the Animal. Take, for instance, the human foot, a solid supporting organ that binds Man as closely to the ground as the supple climbing foot ties the Orang to life in the trees. But, whereas with the Animals those specializations are their inherent means of adaptation, developed from their scheme of body-compulsion, in the case of Man they have grown merely as an aid to his use of tools, and are part of the synthesis of artificial adaptation.

Another suggestion has been advanced to the effect that it was mainly in *compensation* for physical deficiencies that Man has taken to artificial tools. There are, in fact, a great many instances of artificial devices being employed for the purpose of compensation—as when we wear glasses to compensate for defective eyesight, or use crutches to compensate for the loss of a leg, or have insulin injections to compensate for the failure of the pancreas gland, or feed the baby with the bottle to compensate for the lack of mother's milk, and so on. Yet, here again, the essential point is that Man strives to make himself free from body-compulsion by artificial devices, and it was indeed only after Technology had reached a comparatively high degree of perfection that he became able to devise effective compensational adjustments even for such shortcomings as congenital and traumatic defects, deformities, disease, short-sightedness, etc. Thus, while a mammalian Animal that was born a cripple, or had no mother to feed it, or had lost a vital limb or the keenness of its senses, would have no chance of survival, Man, thanks to his advanced Technology, is no longer endangered by such defects. This should not, however, be taken to mean that from his very beginnings Man took to artificial tools as compensation for physical shortcomings, as though his "nakedness" would have compelled him to "invent" the tools. From the standpoint of this book the very opposite assumption holds good:

his loss of natural equipment was the result rather than the cause
of his taking to tools.

When we use a tool we bring the principle of body-liberation
into play, no matter how small be the extent of body-liberation
demanded. Nor does it make any essential difference in which
way we seek and achieve body-liberation; whether, for instance,
in cold winter days we keep ourselves warm by means of a
woollen cloth or a fur-coat, a coal fire, a hot-water bottle, or an
electric blanket. In either case it is an artificial contrivance, a
"tool", that we are using for adaptation to environment *instead
of* effectuating adaptation by means of our body.

With mechanically operated tools, since they require less
"attendance", the extent of body-liberation is naturally greater
than with tools operated by our own organs, and for this reason
there is nowadays a growing trend towards automation for the
purpose of higher degrees of body-liberation. Indeed, the dream
of the modern engineer is a machine running fully automatically,
in which case body-liberation would be complete.

This general tendency to intensify the efforts for obtaining
greater variety, range, and higher perfection of tools and
machines is clearly evinced by the fact that day by day new
"labour-saving" devices are put on the market. All this attests to
a deep-rooted urge in Man stirring incessantly and irresistibly his
spirit to explore and utilize any possible material and power-
source for the sake of greater mechanical efficiency—an innate
urge that can be traced through the whole span of human
history. For, although the modern advance of Technology has
rightly been looked upon as the keynote of our age, the art of
tool-making is in itself no privilege of our time, but a very ancient
pursuit of Man, indeed as old as Man himself. Palaeontological
records furnish reliable evidence of an unbroken sequence of slow
but steady progress in the technique of tool-fashioning, and
strongly indicate that the cultivation of artificial tools has, from
his earliest days, been a matter of supreme concern to Man.

Evidence thus goes to show that present Technology has its
roots deep in the fertile soil of the Stone Age, and Man of today

is but carrying on the enterprise that was started and developed by his remotest ancestors. Once the tool had come into the possession of Man, he himself was possessed by it, and he put into it all his soul and strength. But what early Man with his primitive flints had so arduously striven for and had yet achieved in so small a degree—liberation from body-compulsion by means of artificial tools—present Man enjoys to the full by the ever-increasing wealth and perfection of his technical devices.

CHAPTER 7

Speech

IT MIGHT be argued that the principle of body-liberation may well stand for the circumscribed zone of Technology and its *material* tools, machines, and any kind of mechanical device, which by their extra-bodily nature would readily lend themselves to the adjustment of Man to his surroundings and thereby set him free from the natural limitations and restrictions of his body; but that Technology would not stand for the totality of civilized life, especially regarding its *spiritual* sphere which comprises such prominent fields of human activities as science, philosophy, morals, and the arts. There can, of course, be no doubt that the spiritual sphere of Man is not only an integral and most distinctive part of modern civilized life, but is even inseparable from modern technical progress itself which would not have risen to its present height without the laws, conditions, and structures of the physical world being explored by scientific research. It is due to this supremacy of the human mind that the spiritual achievements of Man are frequently more highly esteemed than all his prodigious technical advances. If therefore the principle of body-liberation is in fact the basic principle of human evolution, then we should expect it to hold equally good for both the technical and the spiritual sphere of Man.

At first sight it may seem difficult to bring the *material* tools of Technology into line with the *immaterial* manifestations of the mind. It is indeed over this problem that many a student of biology, when searching for the dynamic principle of human evolution, has come to grief. And yet, if we are right in conceiving human civilized life as the natural outcome of a uniform evolutionary process, we must expect that the same principle of body-liberation that we found at the root of Man's technical achievements must lie also at the root of his equally unique mental achievements.

Such a supposition should encourage us to venture upon the novel task of reviewing the mental attainments of Man in the *perspective of tool-use*—a task which has not yet been undertaken methodically, but which should be full of promise after we have seen that the material of which a tool is made, although it will largely determine the specific performance of the tool, will never qualify the tool in its essence as a "tool"—that is, as an artificial means made and used for the ends of body-liberation. Supposing Man has in his search for new material also seized upon "mental" material with which to build yet another category of tools which we can term *mental tools*, then the principle of body-liberation would also be operating in his mental sphere.

Our inquiry may conveniently begin with an exposition of Language as presenting the genuine vehicle of thought. A good approach to what Language actually is in its nature and function we may find when we read a book, say, a travel book carrying us in our mind to a foreign country that we have never seen. Yet the author's graphic account of the country, its geological and climatic conditions, its mountains and rivers, flora and fauna, peoples and customs, buildings and arts, brings its scenery and character so vividly to our mind that we become intimately familiar with it and would immediately recognize it should we ourselves happen to travel there. Language, then, is apt and designed to convey knowledge of facts, conditions, circumstances, or indeed of any object, situation, and happening in the world, without the need for obtaining that knowledge by our own perception. This is so self-evident that we do not even suspect this source of information of being a phenomenon of a rather strange and mysterious kind. When we read our newspapers, or "listen in", to learn the news of the day we may perhaps be deeply impressed by the wonders of modern printing and broadcasting; but the enrichment of our knowledge by mere Language would appear to be the most natural thing in the world. And yet it is anything but "natural".

Nature has equipped her creatures with a system of sense-organs enabling them to obtain an appropriate knowledge of the conditions and circumstances of their surroundings. Some vague and indirect knowledge they may also gather from the behaviour and inarticulate cries of their fellow creatures; but any concrete information of what is actually before and around them can only be acquired through the channels of their sense-organs—that is, by their own experience. If this be so with the Animal, it should equally be so with Man. That is to say, for Man, too, the "natural" way of getting knowledge of the outer world is through his senses and his own experience. Therefore when Language allows him to obtain such information without calling upon his personal experience it must appear to be a quite abnormal condition, and the question arises how best to account for this extra-ordinary function of Language.

In its widest sense, Language is a means of communication, and it achieves its ends through the intermediary of its con-stituent elements, the "words", each of which has a special refer-ence to a definite object. In highly developed languages a single word may denote more than one object, most words having here become abstract, or "concepts". However, since Language has not changed its original function of conveying knowledge, there is in this respect no difference between concrete and abstract Language, only that abstract Language has gained in range, precision, and versatility.

In considering first concrete Language, we start from the plain fact that the concrete word is a definite sign denoting a definite object. So when I speak of the rose on the table, the word "rose" denotes, of all flowers, this particular flower on the table. The emphasis here lies on the provision that the word only *denotes* its object. Neither is the word the object itself nor a copy of it, nor does it provide the actual perception of the object, but it is simply a conventional sign which, through its close association with its object, is apt to evoke in our mind, if not the perception, yet the distinct *image* of its object. Thus, if we talk of an absent friend, we do not actually see him, and yet his image will stand distinctly before our mind's eye.

In ordinary life, for everyday needs and ends, this specific function of Language must be of far-reaching consequence. Here indeed the mere image of an object is for practical purposes almost as effective as its actual perception. Whether, for instance, I see with my own eyes that my house is on fire, or I am informed by verbal message, in either case I obtain definite knowledge of that incident and can take appropriate measures.

It follows that Language, through its specific function of calling up the object's image before our mind, *releases* us practically from the compulsion of self-experienced knowledge. This surely is an act of *body-liberation* and suggestive of a close relation of Language to the human principle of evolution. The more so, as through the instrumentality of Language, we are able to obtain knowledge of every thing and event all over the world, whether near or far, present or past. Whereas when resorting to our natural *perceiving* means, the sense-organs, our knowledge is restricted to those objects only that are in their narrow reach.

Language, then, appears to set Man free from the compulsion and the natural limitations of his sense-organs. Hence, the *words which constitute Language must fall into line with the material tools of Technology*: that is to say, they are essentially extra-bodily, or artificial, means made and used for the ends of body-liberation.

At first sight, this may seem a rather bold statement. Do not the words, since they are being *spoken*, clearly indicate that they are a *function* of our speech-organs? One would indeed think that they cannot be anything else than such function, and that the "function-theory" of Language is rightly generally accepted. In textbooks on physiology we may find Language analysed by a detailed description of the mechanism of how the vowels and the consonants, as the components of the words, are formed by the "co-ordinate action of certain organs"—mouth, tongue, larynx, etc.; and in textbooks on psychology, Language may be defined as a *"system of certain bodily actions* used for the purpose of carrying on and expressing thought".

There can, of course, be no denial of the obvious fact that the act of *speaking* a word is, as such, closely connected with the speech-organs, and is a true function of them. But this is not in question here. Our point is that speaking, in the sense of producing the sounds constituting the words, is essentially a bodily function, but *Language is not.* When we speak a word, this function of our speech-organs is neither the word itself, nor does it create the word. Surely, the word has long been in existence and is well known to us. It has, as it were, during the time of its existence been at our disposal and ready for use; and when we speak it we do so for the purpose of *operating* it, that is to say, putting the word's potential power (of imaging its object) into actual effect.

Oddly enough, yet most logically, there is a distinct analogy between the word and the technical tool. When we swing a hammer to drive a nail into the wall, this is clearly a bodily function. Yet this function is neither the hammer itself, nor does it create the hammer. The hammer is already in existence and ready for use, and our bodily exertion only serves to operate the hammer, namely, to put its potential power into actual work. The analogy goes even further. Hand and arm suggest themselves as the natural and original organs for working the hammer. Still, as it is only for operating the tool that we avail ourselves of the function of these organs, the same purpose may be achieved by other means, even mechanical ones. Likewise the speech-organs suggest themselves as the natural and original organs for operating the words. But here again the same purpose may be achieved by other means; and in point of fact, we can operate the words as well by our hands, such as by writing, or by the deaf-and-dumb gesticulating signs, or by no organ function at all but by mechanical means, such as printing.

The mere fact that the words must not necessarily be *spoken* but can also be worked in other ways should in itself suffice to cut the ground from under the "function-theory" of Language.

We may, however, add some further proof in favour of our own "tool-theory" of Language.

1. If Language were a function of our speech-organs, there could be only one language, within the whole of Mankind, in view of the fact that these organs are uniformly identical in structure, and hence in function, among all Men. There are, however, hundreds of different languages in existence.

2. With his one set of speech-organs, individual Man should be expected to speak one language only; he is, however, able to speak several different languages. He should also be expected to speak by nature his own "mother tongue". But if an English child were from birth reared among African natives, he would quite naturally come to speak the language of those people, although in his physique he would develop all the features characteristic of his own race.

3. Any organ-function being inseparately attached to its special bodily organ, it would not be possible for a function to be passed on to another person. Language, however, has to be "learnt"—that is, acquired from others as a ready-made device of expression. A child, therefore, in spite of his normally developed and hence normally functioning speech-organs, would not learn to speak at all if there were not somebody to impart Language to him.

4. Since organ-function ceases after death, Language, if it were a function of the speech-organs, must inevitably come to an end too. The fact is, however, that Language may even survive the death of a whole nation and after thousands of years be "excavated" along with other remains as a true relic of the people that once spoke it.

The inclusion of a mental element (the mental association with its object) might lead to the other question as to whether the word, if not a function of the speech-organs, might possibly be a function of the brain. Is it not a fact that, when we form, or learn, a word, our mind is actively engaged in establishing the association between sign and object, and, when we use the word, our mind again is active in eliciting the object's image? This is quite true, yet it does not make the word a mental function any more than forming and swinging a hammer (all done by our hands) makes the hammer a bodily function.

If we distinguish strictly between *forming* the word, which involves mental function, and the *completed* word in which this mental association is firmly anchored to its perceptible sign, precipitated as it were and enshrined in the sign, this somewhat confused situation will clear up at once. We now realize that forming the word goes with mental function, but the completed word, far from being a "function", is a *self-existing* device ready to be used by anyone, which, however, owing to its mental core, calls upon the exertion of our mind for putting it into operation.

Since the specific performance of a tool depends largely on its intrinsic material, it is the mental core of the word which performs the specific function of calling the object's image before our mind; whilst the word-sign, spoken or written, is merely the perceptible carrier of the mental core. For this reason it is irrelevant whether we speak this or that language, or express the sign by this or that letter, figure, symbol, or gesture; for in any case it is the same specific performance that we want, and obtain, from the sign. Similarly, in the case of a wooden mallet, its "soft" performance remains the same, from whatever trees its specific material, namely wood, has been taken.

If the word is conceived as an artificial extra-bodily tool, we will at once understand the otherwise unintelligible fact that Language, like Technology, constitutes a domain of its own which grows and develops on its own account and by its own principles and dynamics, and like all tools will go down to posterity. The *Oxford Dictionary* is as much a display of the linguistic development of the English language as is the Motor Show for the high technical development of the automobile.

That there is such a close analogy between Technology and Language lies in the fact that it is the same principle that rests at the root of either section. This principle when giving birth to Man started with materials taken from surrounding nature, such as stone and wood, and later on, with the creation of Language, proceeded to draw also upon the mental properties of Man as an additional and prolific source of formative material.

When Man first came to speak we do not know. *Neanderthal Man* of the last Ice Age, although still lacking the characteristic

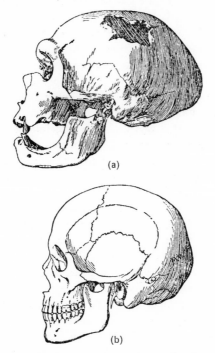

(a)

(b)

FIG. 6. A Neanderthal skull (*a*) compared with a modern human skull (*b*). (By permission of the Trustees of the British Museum (Natural History).)

chin formation, may be taken for certain, by virtue of his advanced brain development, to have been in possession of language. Even *Peking Man*, almost half a million years ago, since he made stone-tools and used fire, may have been possessed of some primitive kind of language. Most probably, speech is still a much older human acquisition.

When early Man proceeded from bodily to extra-bodily defence, he must soon have found himself caught up in an entirely new situation in which his survival essentially became dependent upon his artificial tools and his skill and alertness in using them. In those emergencies which necessitated close co-operation with his co-fellows to ward off dangerous attackers, the old means of communication, such as warning cries and gestures, were no longer adequate and so were effectively complemented by other sounds referring, and understood to be referring, to special vital things or circumstances. To this point we shall come back in Chapter 11.

The close connection between the systematic use of artificial tools and the emergence of primitive word-sounds is consistent with our thesis that it was the same principle of body-liberation which evolved the technical tool, and which also gave birth to the *word* as a new artificial device to free Man, by its infinitely superior powers of communication, from the limited range of his sense-organs. The rise of speech from the first crude word-sounds to primitive yet articulate word-signs, such as are still observable with some backward human races, and from there to the brilliance and versatility of modern civilized languages again bears out our theory that the principle of body-liberation is in fact the evolutionary principle of Mankind.

Accordingly we have reason to suspect that those organs eliminated, or neglected, by the advent of Language have been subjected to *retrogressive* development, and that those organs engaged in the service of Language have been subjected to *progressive* changes.

Since it is perception and self-experienced knowledge that is substituted by Language, we must fasten on to the sense-organs for retrogressive features, and it would indeed seem that with civilized Man these organs, when compared with those of Animals or even of primitive human races, fall far short of sensory acuity. To the Animal such a serious impairment of adaptive sensitivity would be extremely detrimental. So far as Man is concerned such deterioration of the senses does not in the least endanger his survival, thanks to his use of Language which furnishes him with requisite knowledge of the outer world in an infinitely more effective way.

As regards *progressive* development, the most prominent feature, apart from specific modification of the speech-organs, is the enormous expansion and elaboration of the human brain, which in the first place may partly have been due to the new and exigent demands made upon it by the change-over to the extra-bodily style of life, but also must be attributed, perhaps even to a greater degree, to the powerful stimulus that Language exercised upon it by its function of expressing and conveying experiences with definite precision; the more so, when with the higher development of Language a secure foundation was laid for carrying on long trains of thought.

Here again the retrogressive and progressive features combine to make up the specific structure of the human frame.

CHAPTER 8

Reason

OUR analysis of Language has so far been confined to *concrete* words denoting single objects. As mentioned before, the Language of civilized Man has long ceased to consist merely of concrete words. Most words have become abstract or *Concepts.* Not even proper names, in spite of their exclusive reference to individual persons or things, have been saved from this generalizing tendency, and the name of Croesus, king of Lydia, may well stand for any man of great wealth. The biological explanation of this sweeping shift from the concrete to the abstract lies in the observation that a language composed of concrete words only, as still to be found with some primitive human races, is by its verbose circumlocution much too clumsy to keep pace with the infinite number and variety of things, conditions, and happenings in the world of modern Man. On the other hand, the *abstract* word, with its firm grip upon the totality of things, allows Language to rise to a new level of expression in which the exact communication of any and every experience and thought becomes a practicable routine. Judging from the absence, or scarcity, of abstract words in primitive languages we may look upon abstract Language as an advanced developmental stage of communication, yet one which greatly differs from its preceding "nonabstract" stage in that in it a novel and most significant faculty of Man makes itself manifest—*Reason*.

That Man is rational and the Animal is not is an old belief. If this belief is correct, we must expect Reason, as a distinctively human possession, to bear again a close relationship to the principle of body-liberation. It is therefore incumbent upon us to define what is implied by the distinctive mental faculty called "Reason". Modern use of Language tends to interpret Reason on the grounds of "drawing inferences", and since Animals are also credited with the ability of drawing inferences, the old-established distinction between the human mind and the animal mind was bound to fade away.

For philosophers of old times it was self-evident that the outstanding manifestations of human mind such as mathematics, metaphysics, poetry, etc., unique as they were in themselves and an exclusive gift of Man, would demand its absolute separation from the animal mind. It was the theory of evolution and its postulate that the human mind had continuously evolved from the animal mind that led to the modern tendency to deny Reason its former privilege of being a distinctive and exclusive human faculty. Such continuity of development, it was argued, logically would not admit of drawing any absolute line of discrimination. As the noted psychologist James Ward says in his *Psychological Principles*, 1920:

> At what precise point in this development we agree to say that "proper thought" begins will depend upon how we define thought. And apparently no psychological definition is as yet forthcoming that is not more or less arbitrary, and, for all that, fails to effect any clear demarcation between thought proper and thought in the wider sense.

Locke, to my knowledge, was the first to attend to the critical question as to what it was that actually and specifically distinguished the human mind from the animal mind, and he arrived at the conclusion that the distinction was due to the mental faculty of *abstraction* which Man possessed and the Animal did not.

> We have reason to imagine [he said] that they (the brutes) have not the faculty of abstracting, or making general ideas, since they have no use of words, or any general signs. . . . Therefore I think we may suppose that it is in this that the species of brutes are discriminated from man; and it is that proper difference wherein they are wholly separated, and which at last widens to so vast a distance. . . .

Locke's suggestion was taken up by the German philosopher Schopenhauer. On the evidence that all concepts, from the lowest of experience up to the highest of philosophy, had in *abstraction* their common source of origin, he argued that abstract thought was a faculty of its own order, and essentially different from non-abstract, or perceptual, experience. Accordingly he endowed Man with two separate mental faculties : *perceptual* apprehension of things in their spatial, temporal, and causal relations, or the *faculty of Understanding* which was equally possessed by both Man and Animal; and *abstract* or *conceptual thought*, the *faculty of Reason*, which was distinctive of Man, since the Animal lacked the power of abstraction.

Modern criticism has it that there is no novel mental function observable in the process of abstraction that would justify the splitting up of the human mind into two different orders, the one uniting Man with the Animal, the other separating him. Even the ordinary act of perception would appear to involve abstraction, when the perceived object is singled out, hence abstracted, from the rest of surrounding things.

Modern teaching thus holds that the human mind constitutes an indivisible whole and is different in degree only, not in kind, from the animal mind. In consequence, the case "Understanding *versus* Reason" no longer arises, and the term "Reason" has practically disappeared from modern textbooks on psychology, just as if there had never been that noble mental faculty of Man, manifested in his works of science, philosophy, ethics, and the arts, whereby in the words of Locke, "Man is supposed to be distinguished from the beasts and wherein it is evident that he surpasses them."

Such disregard for Man's old-established possession of the faculty of Reason would seem to be a most unsatisfactory state of affairs and demands fresh inquiry. Even if it be that Man cannot boast of any such mental function that would readily separate the human mind from the animal mind, it cannot be denied that there is something exceptional about his "faculty of abstracting, or making general ideas". Suppose Man contrived to make some novel use of his old mental functions by building upon them a new mental device, similar to his previous achievements of using the functions of his hands to fashion stones, and the functions of his vocal organs to fashion words. In this case there would be no new mental function required, and yet an entirely new pattern would have come into existence.

This, surely, is not an easy inquiry, considering the tangled ways of our mental activities, and in order to decide the crucial question as to what Concepts are essentially and how they come about by abstraction, it will be necessary to consider briefly the three categories of ideas.

If I glance at the rose on the table and thus *perceive* the rose, the idea of that rose arises quite spontaneously, that is, *passively* in my mind, and whenever I happen to be reminded of that rose, its distinct *image* will at once revive in my consciousness. Such ideas, prompted by, and referring to, perceived individual objects, are called *Individual Ideas*.

If it is not a single rose that I perceive on the table, but a bunch of roses all of the same type and colour—roses, that is, which are strikingly *alike* in appearance—again an idea of the rose will *passively* form in my mind; yet one which refers to, and is representative of, that whole type or class of roses hence is called a *Class Idea*. Because of the overwhelming likeness of its objects, which makes one rose look exactly like the other, the Class Idea is still associated with quite a distinct *image* of that special type of similar objects.

If the roses on the table are no longer of the same typical appearance but differ among themselves in shape, colour, fragrance, etc., they still are "roses"; if there are also carnations and tulips among them, the bunch is no longer one of "roses" but still one of "flowers"; and if fern and aspen are added, it is no longer a bunch of "flowers", but still one of "plants". Here we are again confronted with a number of objects which, however, are not overwhelmingly alike, but resemble one another only in some certain particulars. Thus the roses may differ in shape, colour, and other details, yet have in common such "diagnostic" features as the thorny stalk, the shape of the leaves, and the terminal growth, general shape, arrangement, and fragrance of the flower. It is on these features that they have *in common*, that is, on their distinct *similarities*, that despite all the gross dissimilarities that separate them we come again to form a distinct idea of the rose, and since all roses of whichever variety, colour, or growth have practically the "diagnostic" features in common, this idea of a rose must be representative of any and every rose in the world, and acccordingly is called a *Universal Idea*, or *Concept*.

There is then a profound difference as against the Individual and the Class Ideas in that the Universal Ideas, or Concepts, do not arise spontaneously in our mind but call for a good deal of *mental activity*. Before us is an indefinite number of objects united by some common mark, or marks, but otherwise throughout unlike; therefore it can only be by arbitrarily ignoring their conspicuous dissimilarities and at the same time, as arbitrarily, concentrating upon their common features that we form their Concept. Thus we form the Concept "White" by singling out, and attending to, the common colour only of otherwise totally differing objects such as alabaster, snow, swan, milk, chalk. This *active* process of singling out, and attending to, the mark common to otherwise throughout unlike objects, in order to form a Concept, is called *abstraction*—a term which literally means "taking away", namely, the marks of difference.

The full extent of mental activity required for forming Concepts is best realized when we consider that by ordinary perception we apprehend things in the world just as they naturally present themselves to our senses—namely, as whole, complex things of definite shape, composition, colour or quality, whilst when we form a Concept by picking out their common characteristics, what we are doing is no more nor less than *forcibly breaking up Nature herself*, and the consequence of the natural percepts being split up is that *their images have gone*.

Concepts are *imageless* and for their lack of images can no longer be "seen" but can only be *thought*. A "rose" consisting only of single features common to all roses, that is to say, a rose which is neither moss rose nor tea rose nor musk rose, nor any other rose, neither red nor white nor yellow, neither large nor small, neither in bud nor blossom, but is none and all of them at once, cannot possibly be *imaged* any longer. Concepts, therefore, are of necessity imageless, and for this reason would be most unstable, hence liable to get blurred or lost, had they not found in the words of Language a firm basis to rest upon. It is in fact only through their anchorage in *words* that the Universal Ideas have become what in the strict sense are called *Concepts*.

The question now arises how it is that the Concepts, although being formed on single features only of their basic objects, come to attain so wide a range of universality as to embrace, in effect, the totality of things. With this question we are striking at the very core of a problem that has long since been hotly discussed and yet lost nothing of its profound perplexity—the problem of the nature of Concepts. Are Concepts a mere matter of mental function, or are they of a different order?

As we have seen, Concepts are formed by abstraction, and by this procedure, natural perception is broken up. It is, however, only in our mind that we break up Nature's order, for although Concepts are built upon the common marks only, they must needs grasp, along with the common marks, the objects themselves in which those marks exist. Thus the Concept "rose", by simply keeping to the common marks, holds all and any roses on earth in its grip. Abstraction, accordingly, turns out to be a mental operation of far-reaching consequences, allowing Man to string the vast multitude of things and events, scattered all over the world, like pearls on the thread of their similarities and to link them together in ordered chains.

The full significance of this achievement of mental comprehension of the totality of things is readily understood if again we consider that by Nature, Man is only able to apprehend those things and situations that are within the narrow reach of his sense-organs; and that in his "concrete" Language he contrived to free himself from the compulsion of self-perception which was replaced by images evoked by concrete words. In "abstract" Language he went a step further when he altogether dispensed with imagery, and instead had the whole of the world embraced and made accessible to his mind by imageless Concepts.

Thus in their capacity of freeing Man from the necessity of self-perception, the Concepts again turn out to be "means of liberation from body-compulsion". That is, they are of the order of "tools", artificial, extra-bodily, self-existent entities.

At first sight, considering the obvious mental structure of Concepts, this may seem a rather paradoxical statement. Yet, if here again we strictly discern the *completed* Concepts from the two mental processes of *forming* and *thinking* them, that is, the processes of abstracting and operating the Concepts, the proper nature of the Concepts as "tools" will clearly stand out. Particularly as we do not necessarily form the Concepts ourselves, but may, and usually do, adopt them ready-made, and may not even think them, but have them operated by other means, such as by the use of a calculator or an electronic computer. The independence and self-existence of the Concepts cannot be better demonstrated than by those cases in which we neither form nor think them.

Otherwise a confused situation is bound to arise such as is revealed by the controversy between psychologists and logicians with regard to the nature of the Concepts. Psychologists, pre-eminently concerned with the analysis of the mental activities involved in the processes of forming and thinking the Concepts, were so deeply impressed by the mental construction of the Concepts that they took those mental functions for the Concepts themselves and consequently were at a loss to bring their "psychological conception" of Concepts into line with "what are called Concepts in logic". Logicians, on their part mostly concerned with the completed Concepts, were inclined to take the Concepts as what they actually presented themselves, namely, sturdy things which could be arbitrarily moulded by definition and freely manipulated in syllogisms. The obvious fact that all mental processes were of a *subjective* kind, varying from experience to experience and from individual to individual, whilst the Concepts were of an *objective* kind, independent of, and greatly surpassing, individual experience, even led some logicians, such as Bolzano and Meinong, to propose an "object-theory" of Concepts in which the Concepts were plainly taken as *"things"*. In this context it should be mentioned that among students of psychology an effort is being made to restore the lost contact with "what are called Concepts in logic". So we read in James Ward's book (*loc. cit.*) that it is "the characteristic of every completed concept to be a fixed and independent whole, as it were, crystallized out of the still-fluent matrix of ideas".

That Concepts are indeed independent things is clearly revealed by the syllogisms of logic where they show themselves capable of being summed up and subtracted like the balls of an abacus, hence one rightly speaks of the "algebra of logic". Concepts may even serve as "basic objects" for new Concepts, in which case they actually stand in line with other solid basic objects from which they themselves were abstracted. Most of the higher Concepts are thus formed by abstraction from lower Concepts. Moreover, Concepts are apt to be conveyed from one individual to another, and even illiterate people may in this way learn a highly abstract language such as English. When this observation was first made with African Pygmies, psychologists were astonished, since brain and intellect could not have developed so considerably in so short a time. If, however, we take the view that Concepts are transferable things, or tools, it will no longer be perplexing that primitive people may learn abstract language as quickly as they learn the use of other tools imparted to them ready-made, such as a rifle, motor-car, or telephone, although in their native country they were still used to arrows, ox-carts and alarm-drums.

As opposed to the narrow range of concrete words, the Concepts represent the more comprehensive tool-form, wherefore it is not surprising that Language, in its natural course of development, has changed from the concrete to the abstract. Yet it is not only the promotion of Language that is to the credit of the Concepts. They have also on their own behalf expanded over a wide sphere of outstanding activities, such as is exemplified by the systems and principles applied in science, mathematics, physics, and philosophy. It is in this very sphere of cultural life, in which the Concepts play a dominant role, that Man's most brilliant gift comes to the fore—*Reason*.

What, then, is "Reason"?

If we follow our theory regarding the "tool-nature" of Concepts, we have first of all to exclude those mental functions which are engaged in the acts of perceiving and imaging, and of drawing *perceptual* conclusions. Whether or not that great mental faculty of *perceptual apprehension* is called by the appropriate name of "Understanding" is of no account. What alone matters is that the perceptual apprehension of things is not confused with the *conceptual comprehension* of the world, which is due to the faculty of Concepts, and it is to this distinctively human faculty, based on the principle of body-liberation, that the term "Reason" should be reserved.

As a "tool-faculty", Reason stands in line with the two other tool-faculties, Technology and Language; hence its analogy with them may help to elucidate the correlation between Reason and mental function.

Hammer, word, concept—all are artificial tools and originally operated by means of special body-function. The hammer is a constituent part of Technology, and when we operate it, this is done by the function of our hand. Yet, when we *use* a hammer, that is, when we work *with* a hammer (instead of doing the work with our bare hands), Technology is brought into play. So is the word a constituent part of Language, and when we speak it, this is done by the function of our vocal organs. Yet, when we work *with* words, that is, when we speak *in* words (instead of resorting to inarticulate cries), Language is brought into play. So the Concept is a constituent part of Reason, and when we think a Concept, this is done by the function of the brain. Yet, when we operate *with* Concepts, that is, when we think *in* Concepts (instead of resorting to perceptual apprehension), Reason is brought into play.

Reason, then, may appropriately be termed the "faculty of *thinking in Concepts*", or, with Schopenhauer, the "faculty of *abstract or conceptual thought*". Accordingly, the term "reasoning" in its strict sense means drawing *abstract* conclusions, in contradistinction to *perceptual* conclusions which are derived from the simple functions of comparing and judging the objects of our *perception*, and are, as we shall later see, shared by Man and Animal alike.

The absence, or scarcity, of abstract words in primitive languages strongly suggests that, at its start, Language was still void of conceptual elements. It means that Man had long been in the possession of Language before he came to contrive Concepts. The very first formation of Concepts may have occurred in such a way that for the sake of easier and speedier communication the originally concrete words gradually assumed a more generalized connotation, and if this were so, the Concepts would virtually have grown out of Language for Language's own perfection.

With the first formation of Concepts, when the foundation was laid for the development of Reason as a third tool-faculty, the same creative principle of body-liberation, which had already manifested itself so impressively in the realms of Technology and concrete Language, took another important step upwards to deepen and widen its operational basis—a new step but one intrinsically coherent with the previous ones. Human evolution, as a whole, can indeed not be properly understood if we do not think of it in terms of a homogeneous process in which the same principle worked itself out in raising Technology, Language, and Reason; the one stage necessitating, for its own development, the next one.

With the advent of Reason the "world of senses" is left behind and the door thrown open to a new world, that of "ideas". All-comprehensive knowledge of things and conditions in the world will come within the reach of Man and enable him to set his Technology upon the groundwork of scientific principles; Language will become a perfect instrument for the promotion of communication and thought; self-reflection and abstract motives will come to play an essential part in human affairs and lift

D

human life upon a higher plane of existence. Thus the advent and rise of Reason shows again the principle of body-liberation in the light of the evolutionary principle of Mankind.

As a tool-faculty, Reason must also be expected to have exercised a twofold effect upon the physical development of Man. With our still scanty knowledge of the intricate nature and function of the various areas of the brain it is, however, difficult to make a definite statement in this respect. One theory is that with civilized Man the number and power of instinct has steadily been decreasing. This, then, would be a *retrogressive* feature which perfectly fits in with our assumption that with abstract ideas and motives gaining influence upon the conduct of human life, the formerly all-dominant instincts—tokens of the principle of body-compulsion—were bound to lose ground when they were superseded by conscious volition. The anatomical "seat" of the instincts is supposed to be localized in the so-called inter-brain, which is the oldest part of the brain and is, characteristically, far bulkier in Animals than in Man. Against this regression, Man can boast of a considerable development of the frontal area of his brain, the "seat of thought"—a *progressive* feature consistent with the rise of Reason, as also is the enormous increase in the number of certain nerve cells, the "neurones" in the grey cortex of the human brain. Again we find here retrogressive and progressive features intermingled in the physical make-up of Man.

CHAPTER 9

Science, Morals, and Aesthetics

So LONG as abstract thought is merely concerned with enlarging the volume of our knowledge, and serves only to promote our ability of coping more efficiently with the varying conditions of the surrounding world, our conduct of life is still apt to be determined by innate impulses and desires, hence it is in this respect not different essentially from the inborn egotistic behaviour of the Animal. If, however, we proceed to form idealistic concepts and allow abstract motives to act upon our conduct of life in such a way as to supersede those egocentric urges, then by such "trans-subjective" behaviour we actually transcend the prescribed zone of self-interest, and thus deliberately sever the natural bonds which connect life intrinsically with the exigent conditions on earth.

This may seem to be a proposition of little, if any, probability. Surely there cannot be the least chance of survival for anyone who would try to escape from the grim pressure of the struggle for existence by simply ignoring its exigencies instead of devoting all possible effort and thought to the supreme task of the preservation of individual life and, indirectly, the maintenance of the species. So vitally imperative is this duty of self-preservation that Nature has secured its strict fulfilment by implanting irresistible impulses and desires in her creatures.

Yet in the spheres of Science, Morals, and Aesthetics we note that Man has created a situation in which idealistic motives are at work to counteract Nature's urgent demands and to override those innate impulses and desires. That his unegotistic attitude may be harmoniously woven into the texture of his evolutionary scheme is a problem that shall occupy us later. At this juncture we are merely concerned with the strange phenomenon that innate forceful mechanisms are deliberately overcome by means of abstract ideas—a definite instance of liberation from body-compulsion which is suggestive of the human principle of evolution here being in play. Indeed, should we find this principle to be also in existence and operation within the *spiritual* domains of civilization, proof would be complete that it has in fact imbued human life in all its essential aspects.

Science, Morals, and Aesthetics are names for vast fields of human activities in which cogitation, volition, and emotion find their supreme expression. Science is said to be inspired and guided by the idea of the *true*, Morals by the idea of the *good*, and Aesthetics by the idea of the *beautiful*. We are not, however, concerned here with the question as to what, in essence, those ideas actually mean and stand for; the less so as their interpretation and evaluation vary considerably with the general cultural *niveau* and the individual level of education. Nor is our inquiry meant to deal with the many other intricate problems besetting those three sublime ideas, their social and emotional values and influences, and their profound metaphysical significances and implications.

It is merely from *biological* points of view that we try to approach the perplexing situation, so as to determine whether there is a genuine relation of Man's spiritual spheres and activities to his principle of evolution. Significantly enough, all three spheres, although each of them is of an order of its own, have in common a distinctive quality which is the basic element on which they are built, and which must bear effectively upon their true essence—the quality of *disinterestedness*, and it is to this fundamental criterion that our inquiry is directed.

In the realm of Science, truth is pursued "for truth's sake" only, that is, regardless as to whether its pursuit may lead to any immediate or remote practical results, or to no result at all; or whether the possible results are to be of any beneficial or even detrimental consequences for Mankind as a whole or for the truth-seeker himself.

The history of mathematics, physics, philosophy, and so on provides many noble examples of such indefatigable and self-denying idealistic search for "pure knowledge" with no realistic ends in view. The same holds true of every line of idealistic endeavour whenever the idea of the true is invoked to guide our mind against the temptations of inborn impulses and desires. From the "objectivity" of the historian who describes the characters and deeds of the prominent personalities and the major events in history *sine ira et studia*, that is, with no partiality for or against; from the "probity" of the judge who passes judgement "without regard to persons", it is only a difference in degree which leads to the scholar who devotes his life and fortune to the pursuit of his scientific research-work, and the self-sacrifice of the martyr who for the sake of his truths would suffer imprisonment, torture, and even death.

In any such case in which the idea of the true is rigidly set *above* and *against* our inner self and its burning needs and emotions, we are actually doing no less than deliberately renouncing Nature's deep-rooted institution of innate impulses and desires. It is clearly in an act of freeing himself from body-compulsion that Man is here struggling, and he achieves his ambition with the aid of an abstract idea. The accomplishment of making himself free from the strong pressure of inborn impulses and desires through the power of his idealism is of course a well-known fact, and has rightly been called the "spiritual freedom of Man". In this respect we see that philosophers have already anticipated the human principle of liberation from body-compulsion, though not yet in the biological sense of an evolutionary principle.

In the realm of Morals it is the idea of the *good* which determines our behaviour, the good act being done "for goodness' sake" only, in compliance with our conception of the good as laid down in solemn moral codes. That is to say, our behaviour is only then truly moral if the good act is done in conscious deference to a principle carrying the idea of good, such as the principle of duty, allegiance, piety and charity, and it is by this criterion of the *consciousness* of our good volition that the moral act stands out against any *instinctive* action enforced by innate impulses and desires, such as the Animal's instinctive care of the young. Moreover, as Kant has emphasized, the moral precepts have to be observed not only consciously but unconditionally, that is, without any element of self-consideration entering into our volition, even against our most vital interests. The wealthy man who parts with all his property in his effort to help the poor, the starving man who shares his last crumb of bread with his comrade, the mother who sacrifices herself for her child, the soldier who dies for his country, are heroic instances of true moral self-denial in which the egotistic impulses, however powerful, are superseded by idealistic volition.

Thus, in the sphere of Morals we again meet with a situation where the pressure of inborn impulses and desires is being over-ridden by an abstract idealistic motive, and here again we note that Man's noble achievement of setting himself free from egocentric body-compulsion has long been known as the "moral freedom of Man".

In Aesthetics it is the abstract idea of the *beautiful* which is placed above and against our needy and greedy self. Kant again has shown that it is on "disinterestedness" that Aesthetics basically rests. We must neither desire nor detest the object of our contemplation, nor bring it into the remotest practical connection with our own life-sphere. Only if entirely detached from any realistic considerations will the object yield to aesthetic contemplation, and as soon as practical associations come in, so will the aesthetic spell fade away. Thus I may be enraptured by the "beautiful" sight of a cornfield teeming with red poppies and blue cornflowers; but if I come to think of the economic damage done by the weeds, the aesthetic spell will at once turn into a wholly unaesthetic realistic calculation.

In Aesthetics, then, we are once more confronted with a situation in which an idealistic idea makes us ignore our practical existential interests—an act of emancipation from body-compulsion widely recognized as the "aesthetic freedom of Man".

Our inquiry was successful in showing that the principle of body-liberation has indeed taken root in Man's higher regions of thinking, acting, and feeling. The outstanding phenomenon of his capacity of controlling his own selfish nature by an idealistic attitude has naturally attracted the philosophers of all times, and has found exalted expression in their interpretations and doctrines relating to a higher order of human existence and destiny. However, from biological points of view, we are bound to look upon the miraculous phenomenon of Man's spiritual freedom as a natural growth arising from the human principle of body-liberation, in the pursuance of which Man has come to add spiritual, moral, and aesthetic freedom to his previous forms of body-liberation. How far this principle has as yet penetrated into the higher mental spheres of Man depends upon his ability to overcome the powerful pressure of his innate impulses and desires by the growing strength of his Reason. Only if he is able to check his natural egotism by controlling his conduct in reference to those ideas has he reached what is called his spiritual, moral, and aesthetic *maturity*. As long as this stage of maturity has not been attained and "interestedness" not yet wholly superseded by "disinterestedness", so will innate impulses and practical motives still take the lead, as when the mother's care of her child is still governed by the innate maternal instinct, or the good work is being done not "for goodness' sake" only, but in the hope of later reward.

Mankind may still have a long way to go before achieving spiritual maturity, yet the fact that at the root of Man's hard struggle towards this maturity we find his evolutionary principle in operation makes us confident, or at least gives some hope, that in the end he will reach this goal.

CHAPTER 10

Tool-use by Animals

ACCORDING to our theory, a deep antagonism exists between the process of *animal* evolution tending towards the greatest possible degree of *body-adaptation* and the process of *human* evolution tending, conversely, towards the highest possible degree of body liberation.

Such cardinal dichotomy should once for all exclude the animal kingdom from the phenomenon of *civilization* which, as we tried to show, is the direct and distinctive outcome of the human principle of evolution. Since, however, there is a general and unshakeable belief that the gap between Man and the Animal is definitely bridged, if for the one reason that the human stage had developed from the animal stage by smallest gradual steps the one stage merging imperceptibly into the other, and since much evidence has been gathered in support of the doctrine that Man is but a higher developmental degree of the Animal—our theory which claims a "difference in kind" between Man and the Animal, but which so far has been discussed only from the angle of *human* life, should be extended, and indeed put to the test, by an investigation also from the angle of *animal* life. This is the more necessary as it is a well-known fact that Animals are not only well acquainted with the use of tools, but are also able to communicate with one another by special sounds, and are even capable of drawing logical inferences.

Surely there must have been a striking similarity between tool-using Man and tool-using Animal, at least in the first place of human evolution when Man is supposed to have used stones and other objects in their natural state, as he chanced to pick them up from the ground. How, then, could he possibly have differed essentially in his behaviour and in his kind from the Monkeys and Apes who are known to make similar improvised use of stones and other objects?

Yet our theory claims that even his simple use of unfashioned tools indicates the first manifestation of the human principle and so opens up an evolutionary stage fundamentally different from the preceding animal stage.

The tricky problem regarding the true relation of Primeval Man to his animal precursor was so easy to be overlooked, or ignored, because one was preoccupied by the illusion that with respect to the simple use of unfashioned tools it was *self-evident* that there was no essential difference between tool-using Man and tool-using Ape. That Apes and Monkeys are in fact able to make such use of stones and other natural objects as Primeval Man had presumably done in his time is evidenced by a good number of well-authenticated observations which clearly testify to true tool-use in their wild state.

Darwin mentioned an observation made by Wallace who on three occasions saw female Orangs "breaking off branches and the great spiny fruit of the Durian tree, with every appearance of rage, causing such a shower of missiles as effectually kept us from approaching too near the tree". Schimper (after Brehm) saw Baboons in troops descending from the mountains to plunder the fields and when they encountered troops of another species a fight ensued in which large stones were rolled down. Then both species, making a great uproar, rushed furiously against each other. Another instance, experienced by Brehm, describes how Baboons, attacked by firearms, "in return rolled so many stones down the mountains, some as large as a man's head, that the attackers had to beat a hasty retreat".

There is clearly a very humanlike note to be seen in these and other instances of tool-use by Animals, particularly if we think of early Man himself struggling likewise with improvised tools—a forceful simile which readily lent itself to the argument that Man and Ape were, in principle, "doing the same", and that from this "identical case" it was only *in degree* that any further progress in tool-use was made by Man.

This brings us back to our previous comment upon this apparently irrefutable argument when we suspected a gross error to have slipped in (see p. 21). As will now be seen, the conclusion that human evolution is different in degree only from animal evolution and Man but a higher developmental degree of the Animal was indeed *correctly* drawn, but as so often happens, it was derived from a *wrong premiss*.

To make our point perfectly clear, let us in the instance of stone-using Man and stone-using Ape postulate that Man throws his stones not with his bare hands but with the aid of a sling. In this case we would no longer argue that both "are doing the same", for we cannot fail to appreciate the special condition under which the act of stone-throwing was performed, namely, with the aid of an artificial contrivance. Naturally one could raise the objection that with an artificial device in play, the example is no longer valid. This is true enough; still it serves its purpose by warning us against concentrating too strongly upon the isolated act of stone-throwing instead of taking due regard also of its special circumstances and its practical connections with the life-sphere of the stone-thrower himself.

If as another instance we choose the case of Man and Ape "climbing a tree" with no artificial contrivance being used, we still would hesitate to jump to the conclusion that basically "both are doing the same", for we cannot be unaware of the striking divergence in the structure of their hands and feet. This instance again shows that the question of whether Man and Animal are *actually* "doing the same" when *seemingly* acting identically cannot be answered adequately unless full regard be taken of the meaning, function, and significance that their action bears upon their specific conduct of life.

To bring out our point still more precisely we may, as a further instance, take a news-item from our morning papers under the headline "Man kills Man by biting him in the neck". The parallel case taken from the animal realm would read something like this: "Tiger kills Tiger by biting him in the neck". Here again we are confronted, as far as the action is concerned, with two identical cases of killing an adversary by biting him in the neck; yet the idea that Man and Animal are here "doing the same" would never occur to us, our attention being unfailingly drawn to the powerful teeth of the Tiger and the contrasting weakness of human teeth.

What would seem to be for the Tiger a "natural" thing to do, namely, to use his teeth when attacking another Animal, appears for Man to be a most unusual, and indeed "unnatural", event, and it is because this act runs counter to human nature that it makes "news". This reflection upon the proper background of those seemingly parallel, or identical, actions gives us a clue for judging them in their true perspective, and for this we cannot do better than take as a guide the distinctive biological agent that determines the shaping of body and life—the principle of evolution.

With Man, evolution turns upon tool-use as a means of body-liberation. Viewed in this light, his stone-defence as a method of taking to extra-bodily means as weapons in the struggle for existence reveals itself as a true manifestation of the human principle. As far as evidence is available it attests to early Man's effort to improve steadily on tool-use by making his flints more efficient by chipping them into suitable tools with a cutting edge. In addition there is also clear evidence of Man's physical development in the direction of tool-use, such as with reference to upright gait—a development which was plainly opposed to the Ape's promotion of climbing as a "natural" form of adaptation.

This dynamic trend of evolution towards the stabilization and perfection of tool-use suggests that Primeval Man would not just casually, now and then, here and there, have taken to his stones and sticks, but once having set his heart on the use of artificial tools, would tenaciously and consistently have stuck to them, until tool-use grew an established routine, becoming, as it were, his "second" nature, and was automatically passed on, and further developed by his offspring from generation to generation.

Tool-use, it appears, was here *integrated into the evolutionary scheme of Man,* and it is by virtue of its incorporation in his evolutionary mechanism that Primeval Man's simple act of using stones in defence gains its significance and dynamic momentum: a matter of supreme concern in the vital task of securing his survival, a token of his new evolutionary principle coming into operation, and a promising start to his unique career.

With the Animals, on the other hand, it is the principle of body-compulsion, in the sense of bodily adaptation to environment, upon which evolution converges. From this all-dominant principle the Apes and Monkeys make no exception, however much we may admire their considerable aptitude for using stones as tools and weapons.

However, although they are in no way separated from the rest of Animals with regard to the principle of body-compulsion, there is still one distinguishing feature that brings them more closely to Man—their possession of *hands.* It is not because their hands are an integral part of their climbing equipment; but as organs provided with fingers and an opposable thumb they can readily be used, apart from climbing, in quite a number of other ways, such as for picking fruit from trees, and for grasping and using stones and other objects for many purposes. However, the relevant point here is that the natural aptitude of their hands for tool-use, far from increasing in the course of their evolution, has on the contrary signally decreased; the hands have become arched and excessively long, with the exception of the thumb which is much too short in relation to the overlong hands and with some Gibbons even has disappeared altogether.

That of all fingers it was the thumb—so indispensable an organ for the proper and effective use of tools—which was forsaken if not wholly sacrificed by evolution seems clearly to indicate that climbing, that is, body-adaptation, was enforced here in preference to, and at the expense of, tool-use. With this trend of evolution it is in keeping that tool-use seems hardly to have any deeper significance in the life of the Apes. There is indeed no evidence that they would have made any appreciable progress in tool-use, or that they could ever have come to the stage of making tools.

To all appearance they are still using stones, as they may have used them millions of years ago, in the raw state in which they come into their hands.

If again we put the question whether Man and the Ape are "doing the same", when using stones or other objects as tools and weapons, it is with different eyes that we look at our problem. While at first sight the seeming similarity of their actions may have persuaded us into the belief that tool-using Man and tool-using Ape are essentially "identical cases", we now realize that from the angle of their evolutionary schemes there is a profound difference between them.

With Man, tool-use appears to be the very soul and destiny of his evolution; with the Apes, tool-use appears to be directly opposed to their inherent scheme of body-adaptation and for this reason has been uncompromisingly overridden by their evolution. Small wonder that in the case of a real threat to life, when the right means of defence is instinctively chosen, they quickly forget the stones and take to their "natural" weapons—their teeth and their capacity for climbing.

Tool-use, therefore, for the Apes can be but a matter of minor significance, not one of vital urgency, a matter of occasion and opportunity rather than of necessity, and hence admits, to my mind, of only the one interpretation that here the Apes are making the greatest possible use of a function still left to their hands within the narrow limits drawn to them by an adverse development.

Looking out for a parallel case within the human sphere, the instance of "Man bites Man" is a good illustration of such an occasional action outside ordinary human conduct. Evidence shows that the human teeth, particularly the canines, although they would have never been so powerful as those in the large Apes, in the course of evolution have greatly diminished in size and strength, and their present state of retrogression has been attributed to reduced function owing to the use of tools, of fire, and so on. Otherwise, if Man in the course of his evolution had tended to use his teeth as his proper means of attack and defence, he would have grown the large and strong canine teeth of the Baboon and Gorilla. As it is, evolution took with him exactly the opposite direction and left him with teeth of reduced size and diminished strength, and of correspondingly little fighting power. Even so, he may still be able to make some bold use of them, should need arise, and in a close fight might bite off his opponent's nose or ear. This is of course quite an unusual event, and there can be no doubt that, if there were a stick or knife near at hand, Man would have taken to those weapons rather than to his teeth. Neither can it be doubted that in the case of a veritable threat to life, as when attacked by a dangerous beast of prey, he would not for a moment think of defending himself with his teeth, but would instinctively do his utmost to get hold of a knife or rifle, or, at the worst, grasp the nearest large stone or stick as *his* means of defence—just as the Apes and Monkeys, when in the grip of real danger, will instinctively take to their teeth or to climbing as *their* inborn means of defence. Therefore, when we see an Ape defend himself, like Man, with stones, or Man defend himself, like the Animal, with his teeth, we must take those efforts for what they really are, namely, occasional and irrelevant, rather "abnormal", actions resulting from the urgency of making the greatest possible use of still-existing, but by adverse development grossly restricted and hence throughout inadequate potentialities. They have no connection with, and no bearing upon, the principle of evolution. Since it is this principle that alone determines the shaping of body and life and has shaped Man and the Animal on entirely different lines, it follows that what is "normal" for the one, that is concurrent with his principle of evolution, must be "abnormal" for the other and opposed to his principle of evolution. Therefore an Ape who throws stones in self-defence is not any more a "Man" than a Man who defends himself with his teeth is an "Animal". Their proper nature cannot be guessed from some irrelevant "abnormal" action, but is determined by their "normal" inborn disposition as evidencing the true manifestation of the basic principle of their evolution.

Now we are ready to answer the question of whether it is "one and the same" when Primeval Man and the Ape are using stones in defence, and the answer is emphatically in the *negative*. What, for the Ape, is but a functional potentiality strongly opposed to his principle of evolution and hence in its scope and efficiency more and more restricted by the adverse development of his limbs —which therefore can only be an occasional, insignificant, and inadequate effort—is for Primeval Man a vital measure, inspired and promoted by his own principle of evolution, which secures his existence and determines his form of life—a measure of supreme concern and leading him to develop tool-use to its highest degree of perfection and to neglect his natural adaptional equipment which subsequently was apt to deteriorate and thus make his existence increasingly dependent upon his artificial tools.

In a short formula:

> stone-using Man acts *within* and *for* his principle of evolution; stone-using Ape acts *without* and *against* his principle of evolution.

In this light, the fascinating incident of tool-use by Animals shrinks to a matter of minor significance; and so does the remarkable sight of the Ape defending himself with stones. Never, in the case of the Apes, has the principle of body-compulsion been broken up by such casual use of stones, since tool-use is clearly in direct opposition to this principle. With Man alone has tool-use encroached upon, and was integrated into, the scheme of his evolution, thus driving a wedge between him and the animal world.

True, then, that tool-use must always have been a matter of little importance for the Ape himself, yet his simple use of stone and stick was to attain eminent evolutionary significance when it was taken up by Man to serve him as a ready *take-off* for launching upon his separate development. In the concept of evolution it is implied that each new stage is derived from, and developed out of, its preceding stage. Therefore the new human principle could not have come into being and operation unless there existed some previous fitting condition from which to start into existence. This requisite precondition was the hand of the Ape (in its original not yet overspecialized form) and the Ape's ability and practical knowledge of making manifold use of his hands. So what was for the Ape a mere disposition and potentiality, for Man constituted the very basis upon which he established his new kind.

How it came about that from such a pre-existing set of conditions Man may have made his start is the speculative subject of a later chapter. One thing seems certain, that the passage from the animal to the human stage must have been coherent and gradual throughout, so that an imaginary observer might at first not even have noticed any fundamental difference between the two stages. And yet, in spite of all superficial resemblances, it is apparent that with the start of Man the principle of evolution, and through it the whole course of evolution, has radically changed. Although there was no break in the continuity of the universal process of evolution, there was a break in its direction. That is to say, it was not the Ape who continued on in Man, but it was the process of evolution which continued on in its upward movement, and by unfolding a new principle created Man as the representative of a new order of life on earth.

CHAPTER 11

Animal Language

IT IS an old belief that the Animals, like Man, have a language of their own, but that we do not understand them. With the advent of the theory of evolution, this old belief gained new ground; for it was now assumed that human language had gradually and continuously grown out of the simple elements of animal language. As J. Ward (*loc. cit.*) put it: "Though all philological detail is doubtless lost in the obscurity of the remote past, the fact of this gradual advance from natural signs to so-called 'conventional signs' is no longer questioned; and its chief features are tolerably clear."

The passage from the animal to the human stage of language would have been so fluent, the one stage almost imperceptibly merging into the other, that it might not have been possible for an imaginary observer to determine with any accuracy the exact point of time at which animal language ended and human language began. Psychologists, therefore, find it difficult to draw an absolute line of demarcation between the two kinds of language. They are, of course, well aware of the definiteness of connotation in human language and fully appreciate its distinctiveness of articulation. However, all that accomplishment of Language in the communication of experience and thought is not regarded as an essentially new and exclusively human attainment, but is held to be a mere matter of higher development due to the considerable growth of the human mind. The current view (and the majority of modern psychologists are here on the side of Darwin) is that human language, as a higher development of animal language, differs from animal language *in degree only*.

Some students of animal psychology even tried to demonstrate the essential analogy between human and animal language by methodologically analysing and defining the animal sounds and setting up parallels to the corresponding human sounds. The American R. L. Garner was to my knowledge the first to make gramophone recordings of the vocalizations produced by Apes and Monkeys, and from his experiment came to the conclusion that the same sound generally meant the same thing, whilst different sounds were accompanied by different gestures.

However that may be, ordinary and experimental observation leaves no doubt but that Animals are possessed of a certain number of natural sounds which can well be distinguished from one another and which in their different meanings are immediately understood by their fellow creatures. Nor can it be doubted that their utterances convey actual information, and are frequently even of an "intentional character", such as the love-calls, danger-cries, and notes of anger, which are clearly intended to act upon others in such a way as to affect their behaviour.

Altogether, Animals possess in their various sounds, cries, and calls an effective means of intercommunication, and in this respect it would seem that their vocal signs fall into line with human language. There is, however, a significant point of difference. If we inquire what actually it is that is expressed by those animal sounds we find that they reflect the whole scale of momentary emotions, affections, and desires—joy, love, jealousy, anxiety, anger, fear and so on. Observation thus leads us to believe that Animals are in their vocalizations restricted to *conveying the emotion caused by some object*, rather than denoting the object itself which was the cause of the emotion. At bottom, then, their vocal signs reveal themselves as purely *emotional expressions*, that is, outward projections of inner moods, and for this reason have fittingly been called "vocal gestures".

Such qualification does not in the least belittle the remarkable ability of Animals for mutual informative communication. Affections and emotions are particularly apt to be expressed by inarticulate sounds, and not infrequently we find even civilized Man, when in a sudden outburst of joy, surprise, or terror, fall instinctively back upon that elementary form of exclamation.

True, then, that the emotional signs of Animals are virtually void of definite connotation, they are still eminently suited to provide such information as is necessary to meet the simple conditions of animal life. More, however, than such unspecified information the other Animal cannot obtain by these vague kinds of emotional signs, and if it wanted to know what actually was the object that caused the emotional sound it had to find out for itself.

Emotional sounds are commonly spoken of as "natural" cries, which is quite a correct description, since, by their nature, they are but a function of the vocal organs. As such, their number can only be comparatively small and fixed. We may indeed presume that Animals are still using the same few sounds as they did millions of years ago.

Unlike the emotional sounds of Animals, the *word* of human language *denotes the object as such*, so as it appears to our senses rather than to our affections, and irrespective of whether it is pleasant or unpleasant, dangerous or not. The word, in fact, stands for its own sake and achieves its explicitness because it is no longer a natural functional sound but is clothed in a special artificial form and invested with a special connotation. By virtue of their artificiality the number of words, unlike the organically fixed number of emotional sounds, was apt to be increased to any extent, so as to cover every object in the world.

The uniqueness of the articulate word, in denoting the object as such, has been hailed by philosophers and linguists of former times as an irrefutable criterion to separate human language essentially from the natural cries of the Animals. Descartes already argued that the essence of human language was to be found in the *objective* reproduction of external objects, as against the animal signs which were mere expressions of *subjective* emotions and affections, and that this constituted a true difference between human and animal mind.

Modern psychologists, however, with an eye on the theory of

evolution, are no longer in favour of any such differentiation between human and animal language, except a difference in the degree of development.

From the standpoint of this book, the old belief in the distinctness of human language finds strong support. For as we tried to show, it is exactly the denotation of the object, and through it, its mental representation, that makes the word a true means of body-liberation and thereby proves itself as a legitimate offspring of the human principle of evolution. While emotional sounds, although giving some good hint at the cause of the emotion, yet fail to reveal the causative object itself and hence do not release the Animal from the compulsion of self-perception—the word, as an artificial symbol mentally linked with its object, frees Man from that compulsion by its power of evoking the object's image before his mind's eye.

This, then, is our theory, and derives from the fundamental difference between the animal sound and the human word: the *animal sound*, as an outwardly projected emotion, is a mere vocal gesture and as such is of the functional or *bodily* order and hence restricted in number and development; whereas the *human word*, as a man-made means of body-liberation, is a "tool" and as such is of the artificial or *extra-bodily* order and hence unlimited in number and development.

Human language thus stands apart from what is called "animal language", which is no true language at all, since Animals have no "words", hence do not "speak". That Parrots are able to learn some words or phrases and use them in more or less appropriate circumstances does not imply that they "talk". If one Animal might be expected to be able to speak, it would be the Ape who comes nearest to Man in so many features, and indeed has a great variety of vocal signs at his command.

However, the elaborate studies on Chimpanzees by R. M. Yerkes and others have made it clear that those vocal signs are expressions of emotions only, not symbols or names for definite things or actions, not true words. Ordinary experience and scientific research thus converge to show that speech, or language proper, does not exist among Animals—from which observation it follows that in this manifestation of the human principle again the Animal has no share either.

How it was that Man came to form his first primitive word-signs is a matter of conjecture. Some authors believe in an onomatopoeic origin, namely, that the natural sounds produced by, or associated with, certain objects were imitated as a means of reference to those objects. Others hold the view that language may have sprung up in such a way that the originally emotional sounds would gradually have attained a more objective significance when the same emotional sound was consistently referred to the same object that gave rise to the emotion. "For the primitive man", says J. Ward (*loc. cit*), "what he calls out, when he sees a thing, comes back to him as the name of the thing, when he sees it again. Even though altogether subjective in origin, it becomes in the course of repeated experience quite objective in sense."

There is certainly much truth in either theory. Still, it would seem to me that we would miss the most essential point if we failed to recognize the close relation between the origin of language and the peculiar and necessitating circumstances connected with the adopted routine of using artificial tools and weapons. A step of such extraordinary consequence as that of passing from the emotional sound to the naming of objects should not be conceived as just a chance product of evolution, a matter of sheer incident. Rather have we to assume that it was an equally extraordinary cause that urged speech into existence—*a pressing biological need* such as may have arisen when with the established practice of extra-bodily defence Man found himself caught up in a new and menacing situation in which his survival became exclusively and inexorably dependent upon his skilled use and choice of artificial weapons in close co-operation with his fellows. In critical moments, when concerted action was required to ward off or hunt a dangerous beast, he would have felt an acute urge to give the others exact information of what was before them, and by using the same sounds for the same kind of things, beasts, weapons, he would in time automatically have come to connect those particular sounds with their particular objects. The essential point here is that it was in pursuance and in consequence of tool-use that the vital need made itself felt of denoting special things by special sounds, or *names*.

Language thus presents itself in the light of a natural product in the course of human evolution, and at the same time as the fruit of a dynamic principle that began with material tools and now, in support of, and in conformance with, its extra-bodily scheme of existence, gave birth to the *word* as a new tool-form which was to open up a new phase in human history. There was certainly no break in the universal process of evolution when emotional sounds turned into words, but there was obviously a break in its direction when the human principle encroached upon the animal inheritance of vocal sounds so as to use them as a basis for building upon them the new realm of Language.

The old question why Animals do not "speak" has been much discussed since the rise of the theory of evolution, but has not yet found a satisfactory answer. One theory is that their speechlessness is due to their vocal organs not being structurally appropriate for speech. Another theory holds that some Animals may well possess the physical requirements for producing articulate words, yet would not speak because they have "nothing to say". Neither theory, it would seem to me, is on the right track. As to physical fitness, I fail to see why Animals, especially the Apes, so closely akin to Man in their physical make-up, should not be able to produce some rudimentary word-signs since they have managed to produce quite a number of emotional signs. As to mental fitness, I fail to see why Animals, here again the shrewd Apes, should not possess such low grade of sagacity as is required for forming very primitive word-signs, such as denoting familiar things surrounding them, which are of particular interest or concern to them.

Nearer to the point comes L. H. White when he suggests that "it does not occur to the animal to use words as symbols".

Still we would ask: Why does it not occur to them? My answer is, Animals do not speak because they do not feel the *biological need* for forming and using words, since their evolutionary principle is opposed to language as a tool-faculty—just as for the same reason they feel no biological need for fashioning stones into tools.

With Man, the creation of language was *within* the scheme of his evolution, and meant for him new gain in body-liberation, although it was acquired at the expense of sense-acuity.

With the Animals, any weakening of physical fitness would run *counter* to their evolutionary principle, and hence could not be tolerated. Yet the vocal apparatus of the Apes may still be adequate for the production of rudimentary word-signs, just as their hands, despite adverse development, are still being used for the crude manipulation of stones. But if this be so, and even if it be true, as Garner insists, that the Apes make certain approaches to the formation of true words, their use of such near-words would be kept, like their use of stones, within the narrow limits drawn by their principle of evolution, so that to all intents and purposes the integrity of body-adaptation be maintained, only that such approaches, which for the Apes themselves is only a matter of little significance, may again have served Man as a ready precondition from which to start out upon his own new line of evolution.

CHAPTER 12

Animal Mind

THERE is no Reason without Language. Not only are the General Ideas or Concepts imageless and impracticable, and therefore dependent upon the words of Language to find in them the firm hold they need for becoming fit for practical use and development, but it was Language which raised human mind to such a height as was necessary to overcome the difficulty of abstraction, and to form the first primitive Concepts, thereby preparing the ground for the emergence of Reason.

Since Animals lack the word-language, it should be certain that they have no concepts and no capacity for conceptual thought, and hence are totally void of Reason. In former times nobody would have doubted that the gift of Reason was an exclusive prerogative of Man which separated him distinctively from the irrational Animal. Again it was the theory of evolution which in its tendency to bridge the gap from Animal to Man led modern psychologists to reject the old doctrine and to advance instead the opposite theory to the effect that human mind was not different from animal mind except in its higher degree of development.

Darwin himself was of opinion that Animals were well capable of forming abstractions. "When a dog", he argued, "sees another dog at a distance, it is often clear that he perceives that it is a dog in the abstract; for when he gets nearer, his whole manner suddenly changes, if the other dog be a friend." Another instance: "When I say to my terrier, in an eager voice, 'Hi, hi, where is it?', she at once takes it as a sign that something is to be hunted, and generally first looks round quickly and then rushes into the nearest thicket to scent for any game. . . . Now do not these actions clearly show that she had in her mind a general idea or concept that some animal is to be discovered and hunted?" Darwin's suggestion was taken up, on a grand scale, by J. Romanes. According to this author we are not justified in denying the Animals the basic faculty of abstraction, since they are well able to form at least such general ideas as sweet, bitter, hot, cold, good-for-eating, not-good-for-eating, and so on, and it is merely for their lack of word-language that they are incapable of carrying their abstractions into true Concepts, and are only able to form pre-concepts which differ from true Concepts in not being linked up with language—a mere matter of degree.

Against the widespread tendency to interpret objective observations on Animals by way of one's own subjective experience, many psychologists have raised a warning voice that we must not read into the activities of Animals more reflective thought than is really in them, and must not apply to them the human standards of abstract thinking, as if it were the Animal that had made the abstractions, and not Man. Indeed, if we were to grant the Animals the ability of forming and entertaining such abstract concepts as hot, cold, sweet, bitter, good-for-eating, and so on, we would credit them with a much greater amount of abstract thinking than we find in Man himself who as a rule does not take to conceptual ideation when experiencing those simple sensations.

Viewed in their proper perspective, observations on animal behaviour seem to indicate that there is a distinct line drawn to the animal mind—its restriction to the *perceptual* and the *concrete*. That is to say, Animals are well able to apprehend whole objects and situations in their practical relations to given circumstances and immediate interests, and to respond to such concrete perceptions by appropriate action. More, however, than an intelligent adaptation to perceptually and practically grasped concrete conditions is not revealed by their actions. In his book *Mind in Evolution,* 1910, L. T. Hobhouse said: "If we grant the highest animals concrete experience, the power of grasping perceived relations and applying them afresh to new and different circumstances, we grant . . . enough to explain the most intelligent action." Observation and experiment, such as W. Koehler's studies on Chimpanzees, leave no doubt but that Animals are only able to apprehend what is "present before their eyes". Within the circumscribed zone of the concrete and the practical, however, Animals show a keen mental activity by forming not only individual ideas of concrete objects and situations, but also class ideas of similar objects and conditions. As Hobhouse concluded :

> Animals are influenced by similarity of relations. Not that they dissect out the common element which constitutes a class identity. It is rather that they have a concrete perception of the man or animal, house or locality with which they are familiar; and that when they meet another object similar in general character . . . to the first, they know how to deal with it.

The decisive point here seems to me to be that Animals "do not dissect out the common element", but tend to grasp things and situations as they appear to their senses, as natural wholes in their natural context. They do not break up perception, nor do they grasp qualities and relations in isolation from their objects, hence do not think in imageless concepts.

If we discriminate strictly between the *perceptual* ideas on the one hand and the *non-imageable* abstract concepts on the other, it becomes evident that Animals, although they are well disposed to form any kind of *perceptual* ideas, cannot be credited with the capacity of forming *abstract* ideas. Even Man, it would seem, was not capable of conceptual thought until his intellectual level had reached such a height as to allow him to overcome the difficulties of the process of abstraction. To deny the Animals the faculty of Reason does not, however, mean to deny to them the ability for "drawing inferences". There is plenty of evidence that they are well able to do so; only that their lack of conceptual thoughts restricts them to *perceptual* inferences only, that is, concrete inferences drawn from perceptual situations. The term "reasoning" in its strict sense is therefore not applicable to them and should be reserved for the *abstract* inferences of Man, so as not to blur the line of distinction between animal and human mind.

The question why Animals are not possessed of Reason finds its answer historically in their failure to achieve the two essential requirements for the development of Reason—Language and mental adequacy. From a biological point of view the answer again is that Reason, as a true manifestation of the distinctive human principle, is opposed to the evolutionary principle of the Animals. Moreover, Man has acquired his new faculty of Reason at the expense of innate instinctual properties which were superseded by conscious reflections. With the Animal, the instinct powers are its indispensable safeguard in the struggle for survival and any weakening of their infallibility must lead to disastrous consequences. Even Man, though in a much more secure position, may come to harm should his conscious reflections lead him into a precarious state of error, or even of wavering irresolution when an instant decision is needed.

If not Reason, Animals in any case share with Man all those mental functions involved in the processes of apprehending, comparing, and judging the *perceptual* things and situations of their surrounding world—that is, those combined mental functions which are said to constitute the faculty of *Understanding*, and which provide the Animals with all the essentials of knowledge and apprehension necessary to cope successfully with the exigencies of their existence. Man, of course, thanks to his *conceptual* comprehension of things, goes far beyond that narrow perceptual horizon of animal mind. Still, his greater intellectuality is not just a higher degree of animal intellect. Wherever and whenever abstract thinking breaks in, it is Reason that comes into play and with it the human principle of body-liberation which at once sets up a difference in kind. For their lack of Reason, Animals should not be expected to form the abstract ideas of the true, good, and beautiful, and hence should be excluded from the human fields of science, morals, and aesthetics. However, here again efforts have been made to bring them into line with Man, on the persuasive pretension that animal life was by no means without certain approaches, or even "pre-stages", to those human prerogatives.

Nobody would go as far as to contend that Animals would be able to pursue anything like pure science. Man himself was not to rise to scientific maturity until his intellect had grown up to such height as was necessary to overcome natural egotism by a super-individual attitude of mind. However, against the traditional doctrine that science, as an exclusively human institution, separated Man distinctively from the Animal, the objection was raised that the greater part of scientific research had always been, and still was, carried out for the realistic ends of practical application rather than for the idealism of pure knowledge; furthermore, that pure science was preceded by a pre-scientific stage in which observation and investigation were exclusively made for immediate practical ends, and since Animals, too, showed definite signs of "curiosity" and of "inquisitiveness", the conclusion was ready to hand that there could be but a difference in degree between the scientific stage of advanced civilization and the pre-scientific stage of both primitive Man and the Animal.

It can, of course, not be doubted that Animals, like Man, are well capable and desirous of making practical observations and investigations of their own accord, so as to find out whether or not some strange or hidden object is of interest to them, or has any immediate bearing upon their life.

Nor can there be any reasonable doubt as to the assertion that from the stage of pre-scientific investigation a continuous line of development gradually led to the true stage of pure science. As E. Mach has suggested, "observation made at first for the purpose of purely practical ends was later felt to be interesting for its own sake, and so became an end in itself regardless of its practical import". Yet here again, continuity of evolution must not be mistaken for "gradation" in the sense that the new stage was different from the previous one in degree only.

The step from the subjective self-interested behaviour of Animals to the objective idealistic attitude of scientific Man signifies the shift from the animal to the human principle. Therefore, even if it be admitted that the Animal's desire for practical knowledge does not essentially differ from the human pre-scientific stage, yet the actual stage of true science in which thinking is set free from the pressure of immediate needs is given to Man alone.

In the realm of Morals, the tendency towards bridging the gap between Man and Animal has even led to still more pretentious claims. Here it is no longer the "pre-stage", but the moral stage itself that has been granted to the Animal. Thus the fidelity of the Dog has been referred to as being based upon a "moral idea". It would be "utterly arbitrary and unjust", the argument runs, "to deny to the dog the ability of forming moral ideas for the only purpose of reserving the moral sense as a prerogative of Man alone".

Of all the animal "virtues" the most admired one is perhaps the "affectionate maternal devotion to their young" among Birds and Mammals, which indeed closely corresponds to, and sometimes, as in the case of the proverbial "Monkey-love", may even surpass the human mother's care of her baby. This "touching" behaviour of Animals has induced some authors to describe it in terms of "morality" and to apply the same moral significance to both the human and the animal parental care. From a biological point of view, however, it would appear that with the Animals, what is hailed as their "devoted attachment to their progeny" is but a matter of instinct.

The preservation of the species being largely dependent upon successful propagation, provision has been made by Nature in the higher interest of the species that such a vital matter is not left to the Animal's free will, but is forcibly secured by strong innate mechanisms. This is evidenced by the observation that within the same species the mode of propagation and of parental care is among all its members exactly and infallibly the same, whether the brood be supplied with adequate food in advance to live on for a time, as in the case of the Solitary Wasp depositing her eggs in a larva paralysed by her sting, or the young to be fed after birth by the animal mother herself until they are fit to take care of themselves, as in the case of Birds and Mammals.

Since, then, the animal mother's "affectionate devotion" is prompted by the same maternal instinct, it is in principle irrelevant which of the various ways is pursued to achieve the ultimate goal of species-preservation. Those authors who are inclined to take such maternal affection in a "moral sense" should have to praise the Wasp for thoughtfully providing her future brood in advance with food; still more the Swallow for her indefatigable self-devotion to her young, first by hatching and then by feeding and training them. They should therefore blame the Cuckoo for deplorably neglecting her maternal duties when she deposits her egg in another bird's nest, there to be hatched and looked after. Such are the implications from which we cannot escape if we apply moral standards to actions of mere instinct. Morality, we saw, is based upon the conception of the good action deferred to and ruled by a principle carrying the idea of "good". Instinct, on the other hand, is an adaptational equipment of compulsive character which leaves the Animals no free choice of action, but forces upon them a rigid scheme of definite activities.

The distinction, then, between the moral attitude of Man and the instinctive behaviour of the Animal rests on the essential criterion that the good action of Man is done in obedience to, and in conformance with, the idea of "good", and as such is subject to the human principle of body-liberation; while the instinctive action of the Animal is done in obedience to, and for the satisfaction of, innate impulses, and as such is subject to the animal principle of body-compulsion.

Closely connected with the function of propagation is the habit among higher Animals of living in couples during mating time or even for a longer period, a close association which shows a remarkable resemblance to the human form of marriage, and for this reason has again been taken in a "moral" sense. Yet the observation that their mating association is by all members of the same species practised in exactly the same way must strongly suggest that their partnership is actuated by instinct in the higher interest of the species and in close connection with their particular mode of propagation. Therefore we find that "living together in couples" rarely exists in the case of the lower Vertebrates but more frequently with those higher Animals, the Birds and Mammals, which are appropriately and for a longer time linked together in the joint task of bringing up their young.

Another form of association among Animals, known as

"symbiosis", seems to be inspired by the motive of mutual aid. A good instance of such kind of co-operation is the union between the Hermit-crab and the Sea-anemone. The immobile Anemone sits on the back of the Crab, and by being transported by its companion is given the opportunity of obtaining ample food, whilst in exchange it protects the Crab from enemies by its stinging properties. When changing shell, the Crab is said to seize the Anemone with its claw and to plant it on the new shell. Were we to put this touching instance of "affectionate comradeship" in terms of morality, we should have to call it by names of such noble virtues as enduring friendship, faithfulness, devotion, duty, responsibility, gratitude; while in fact their union clearly serves the purely realistic ends of mutual benefit. A similar motive may have led Man himself to create some special forms of artificial symbiosis when he domesticated wild Animals for his own interest. In the first line stands the Dog, the "friend of Man", whose fine behaviour has frequently been compared to, if not placed above, human behaviour. We should, however, bear in mind that the Dog has been the companion of Man since ancient times, and has by education been fully adjusted to the human way of life. At bottom, then, his friendly behaviour is determined, more or less, by training and habit, by his desire for being praised and by his fear of punishment and last but not least, by his clear knowledge of his existence and welfare being entirely dependent upon the mercy of Man—all of which are very realistic motives.

Dismissing all actions based upon instinct, desire, self-interest, or education, we find that on the part of animal behaviour nothing is left that could deservedly be labelled as "morality". The mistake made was to stretch the moral concept so wide as to cover any kind of "regard for others", instead of reserving it for those cases in which the idea of "good" was made the guiding principle of the action. Therefore, if we take "morality" in its strict and proper sense, not only have we to exclude from it the Animal altogether, but primitive Man as well for all the long period during which his actions were still dictated by instinct or egotistic motives. Only when instinctive egotism is deliberately suppressed by an idealistic attitude towards life, and altruism consciously exercised for its own sake, is the stage reached in which the principle of body-liberation has taken root.

The question why Animals are void of morality finds its answer again in the antagonism between the human and the animal principle of evolution, the former abrogating, the latter promoting, the instincts as the guardians of the species. That Animals are lacking in morality does not, however, imply that they are "immoral". Such would indeed be the case in the many instances of animal "cruelty", if Animals were actually possessed of, or credited with, a "moral sense". As it is, Man alone has the privilege and responsibility of being endowed with a moral sense; hence he alone may fail in his moral duties and fall in guilt of "immorality". Animals cannot possibly sin against a virtue that they do not possess. They are *a-moral* rather than immoral.

In the province of Aesthetics, too, attempts have been made to align Animals with Man. Here it is the "sense of beauty" that Animals are alleged to reveal. Darwin himself set the fashion, when he based his principle of sexual selection on the theory that the female Animal was attracted by the "beauty" of the male. From an aesthetic point of view, that is, seen with human eyes, the sight of the male Peacock displaying his pompous plumage is indeed enchanting beyond measure, but the question remains whether Animals apprehend things and sights in the same way as does Man. Experiment seems to show that the colour sense in Animals is somewhat different from that of Man, and we cannot be sure whether the female Peacock does at all perceive the "beauty" of her suitor. But even if it be so that she is in some way or other favourably impressed by his ostentatious appearance, her delight would be essentially of the erotic kind, while aesthetic contemplation is of the very opposite kind with its suppression of any sensuous desire as a prerequisite condition.

The common habit among primitive human races of adorning themselves with ornamental paints and other decorative devices for many purposes including erotic attraction may lead us to believe that there is a continuous line of evolution running from the Animals' erotic display over that of primitive Man to true Aesthetics, such as may have proceeded from simple curiosity to true science, and from instinctive "regard for others" to true morality. Here again the course of evolution changes radically when, with the principle of body-liberation coming into play, interestedness is supplanted by disinterestedness, and again it is the antagonism between the animal and the human principle that bars the Animals from access to the sphere of Aesthetics. Nothing could be more certain than that an Animal looking at things in its surroundings with aesthetic rather than interested eyes would be doomed to early extinction.

CHAPTER 13

The Uniqueness of Man

IN SUMMARIZING the results of our inquiry we should at once mention that its main result, Man's *uniqueness* on earth, is too obvious and significant a characteristic to be overlooked or ignored, and has indeed found full appreciation even by those authors who otherwise strongly advocate the *animal* rank of Man. For them Man is unique because his mental capacities greatly, yet only in degree, transcend the level of the animal mind. For others, Man is unique because his mind essentially transcends the animal mind. From the standpoint taken in this book, Man is unique because, thanks to his unique evolutionary principle, he constitutes an order of his own on earth, outside and beyond the animal order.

The uniqueness of Man manifests itself pre-eminently in the process of *civilization* which, we saw, is the direct outcome of his exceptional evolutionary principle—exceptional because it is novel on the earth and in stark opposition to the dominant principle of animal evolution. According to our theory, the animal principle turns upon the body for its task of adaptation to environment and remains dependent upon, and restricted to, the means of the body, while Man in his evolution follows a principle which by means of extra-bodily or artificial tools releases him from body-compulsion. With Man, therefore, it is no longer the body which develops into specialized forms of adaptation, but it is the artificial tool on which evolution is henceforth focused, whereas the body only follows secondarily this trend of evolution and thereby forfeits its original adaptive outfit.

Man's close association with, and dependence on, artificial

tools is common knowledge. Equally well it is known that Animals do not make tools. This obvious evolutionary antagonism between tool-making Man and non-tool-making Animal points clearly to the uniqueness and separateness of Man. However, the old doctrines of the separate creation and the separate place of Man were with one stroke overthrown as no longer compatible with the theory of evolution. As to the first doctrine, its abrogation must be accepted by everyone who believes in evolution as a continuous process embracing all living beings on earth, including Man. As to the second doctrine, it would be difficult to brush it away, as to all appearance Man does occupy a separate place. Here the point was made that his distinction was only relative, not absolute. Human evolution, it was argued, was basically evolution of mind, and human mind was but a continuation and gradation of animal mind. In this way, the obvious fundamental difference between Man and Animal sank to a mere difference in the degree of development, and there was no separate place any longer in question for Man. He was but a higher developed Animal, a higher developmental degree of the Ape.

The fallacy of this line of argumentation has been exposed in the previous chapters and shall not be repeated here. One aspect, however, still deserves particular emphasis—the unity of the process of human evolution as warranted by the sameness of its underlying principle.

This unity is again an obvious thing. It has been interpreted as deriving from the human mind as the central and unifying agent of human evolution, which interpretation was supported by the demonstrable enlargement and elaboration of the human brain in the course of evolution. There can be no doubt as to human evolution making for brain development; but so did the evolution of all Primates in connection with their habitat in the trees. The anthropoid Apes were already highly intelligent, and Man just followed this trend of development to a still higher level in connection with his new scheme of evolution.

Although brain development was a decisive factor in human evolution it cannot be said to have been its primary cause. Man did not "invent" his artificial tools, words, concepts; they have grown out of his distinctive principle in the natural course of his evolution. When he started upon Technology with his first crude stone weapons his brain was most probably still at, or near, the anthropoid level but gradually developed to a higher level in close correlation to the development of his making and using artificial tools. The successive emergence of Technology, Language, and Reason must not, however, be taken in the sense of single independent evolutionary events of a more or less accidental kind. Only if human evolution is conceived as a uniform and unitary process in which all parts and phases are connected with one another in close interdependence and interaction can we arrive at a proper understanding of that miraculous creature called "Man".

According to our theory, human evolution reveals itself as a process growing naturally and necessarily out of conditions primarily and essentially connected with the use of artificial tools and leading as naturally and inevitably to the development of Language and Reason. The close causal connection between Technology and the two other tool-faculties is still to be seen today in their harmonious co-operation, one with the other, in order to deepen and widen, along with their own development, their common parent principle.

From the sameness of this principle it follows that the unity of human evolution has never been broken, neither by the creation of new tool-forms nor by new levels of mental abilities. The observation that with the development of Reason a new era embarked upon human conduct of life does not yet make Reason the basic and guiding principle of human evolution. In point of fact the material tool of Technology, with which human evolution began, still plays today, on a larger scale, the same dominant role in human life as ever before.

Another characteristic of Man, again too conspicuous to be disregarded, is his possession of *freedom*. Usually, one discerns between the physical freedom of Man, that is, his independence of environment by means of artificial devices, and the spiritual freedom of Man, that is, his release from the pressure of innate instincts and desires by his power of reflective thought. In particular it is his spiritual freedom that is hailed as the essential feature making Man truly "Man". Only in the spheres of his spiritual freedom is he believed to be redeemed from the bondage of the body and lifted up to an idealistic attitude towards life, when he leaves behind him the material word of need, greed, and misery, and enters the sublime world of self-reflection, virtue, and beauty.

From a biological point of view, both physical and spiritual freedom are equally rooted in the same evolutionary principle of body-liberation, and in this context, spiritual freedom appears to be the natural complementary addition to physical freedom, and the ripe fruit of a principle which is still pursuing, on a wider basis and at a higher level, the same direction that it took when it launched Man into existence.

From a metaphysical point of view, a higher value seems to be attached to the attribute of spiritual freedom, on the assumption that its possession allows Man to transcend the boundaries of his earthly existence. This intricate point shall be touched upon in the last chapter of this book. Here we are only concerned with its biological consequences; and in this respect, physical and spiritual freedom—as manifestations of the same basic principle—are standing in one line, the one fertilizing and promoting the other. There can then be no fundamental difference between the successive stages of human evolution, neither with the emergence of a new tool-form, nor with the emergence of a new form of freedom, all of them serving the same ends of enlarging the radius of action of the same original principle.

As this principle has never changed its character, so has Man, in whom it is embedded, never changed his status of being "Man". However, as this principle has risen from primitive beginnings to high development, so has Man travelled along the human road, from the time he opened the gateway to human kind with his crude tools, to the present stage of rich accomplishment. In other words, between Man at a lower level and Man at a higher level of evolution there is no difference in kind, but only one in the degree of development, and Primeval Man, the first of his kind, deserves the name of "Man" as legitimately as all his progeny, prehistoric and historic.

The fact that Mankind is split into a multitude of races at different levels of development does not in the least qualify the issue. For in spite of the manifold varieties which on the surface seem to separate them, their basic unity is evinced by both the fundamental identity of their physical structures and by the fundamental identity of their civilizations. As W. D. Wallis says in his *Introduction to Anthropology*, 1927:

> All the fundamental activities which characterize civilization are found in savagery, usually in a less developed form. In a very real sense, therefore, a primitive culture is a civilization, less complex than is our own, but each has dormant within it all the achievements which mark the superiority of the present age over those which have preceded.

There is in fact no human race existing without being in the possession at least of manufactured tools and of Language, and scientists have been wondering, not so much about the fact that all races are possessed of their own kinds of artificial tools and their own methods of making and using them, but rather that all over the world, even with the most primitive human races, we find Language in existence and even developed along similar lines, though differing in their word-signs. For this ubiquity of Language the German philosopher Herder once gave a good explanation when he argued that "the development of Language was as natural to Man as his nature".

As I would put it, all forms of civilization are basically identical, because it is the same basic principle that works itself out in them; that is to say, they are "natural" to Man because they are inherent in his "nature" as an integration of this principle.

From biological points of view, then, all Men are developed and united by the same evolutionary principle, and however greatly they may differ in their appearances and habits their difference can only be in the degree of developmental progress on the same basic principle.

The recognition of the fundamental unity of Mankind gains profound significance if we consider that the Animals are equally developed upon, and united by, a basic evolutionary principle which, however, is diametrically opposed to that of Man—their principle of body-compulsion. The over-emphasized use of tools by Apes has wrongly been taken to be conclusive evidence of a fundamental identity with the use of tools by Primeval Man. As we pointed out, the antagonism of their evolutionary principles must exclude the Apes, and for that matter any other Animal, from the human line of existence and development. It is only that their ability and casual practice of tool-use may have served Man as a natural take-off for starting upon his new scheme. While in this way the process of evolution continued when the new human principle came into the world, it was no longer the Animal that continued on, and reached in Man a higher level of animal existence. But it was Man who, thanks to his new principle, broke away from the animal order and started upon a scheme of his own, and the Animals continued to develop on their old principle, and for this reason never rose to have more than a rudimentary use of tools.

The sameness, unity, and exclusiveness of the human principle offer a possibility of defining Man on biological lines. Broadly speaking, *Man is of all living beings the exclusive species whose existence and evolution is conditioned and determined by the exclusive principle of body-liberation.*

With this general biological definition which marks Man off distinctively from the Animal, a new approach is made to an old problem. The question as to what it actually was that distinguished Man essentially from the rest of creation and bestowed upon him material power and spiritual greatness has long since exercised the human mind, and there was hardly a prominent feature in civilized life that has not eagerly been seized upon as a convenient expedient for grappling with that problem. Thus Man has been labelled as a political, social, historical, speaking, laughing, rational, aesthetic, philosophical and religious creature. All these definitions have two things in common. First that they are referring only to a particular attribute of Man instead of grasping his whole personality; and second, that they may hold true for civilized Man, but must fall short of conclusiveness if applied to Man of earlier human history.

More comprehensive and hence more appropriate, since it refers to the most distinctive feature common to all Mankind, civilized and uncivilized, historic and prehistoric, is Benjamin Franklin's definition of Man as a *tool-maker*. With this definition he strikes at the very root of the problem. Still it only holds good if the term "tool" is stretched beyond Franklin's terminology so as to cover the material tools of Technology as well as the mental tools of Language and Reason. Even so there is a weak spot in his definition, since it refers to the *means* of the human principle—tool-making and tool-using—rather than to its significant *ends*—body liberation. Preference, therefore, should be given to a definition which turns to the decisive criterion of body-liberation itself; and as the possession of *freedom* is so obvious and distinctive a feature of human life, many attempts have been made in this direction as well.

Above all, it was the fascinating phenomenon that Man was able to set his thought, volition, and feeling free from the pressure of innate instincts and desires that has occupied the imagination of our philosophers. However, instead of searching for the unique principle that must be supposed to lie at the root of such a disinterested attitude to life, they were prejudiced by the idea that human freedom would essentially amount to freedom of will, in the sense of the freedom of moral action. In consequence the problem of human freedom drifted away into the channels of metaphysics where it ended in fruitless controversies over the question of whether or not his freedom of will would enable Man to transcend the laws of Nature.

Although it is true that the possession of freedom was rightly chosen as an exclusive prerogative of Man, differentiating him fundamentally from the animal order, yet what strictly was to be understood by "human freedom" remained obscure.

According to the theory advanced in this book, all freedom of

Man, physical and spiritual, emanates from one and the same source—the principle of body-liberation. That is to say, his liberation from the compulsion of bodily organs, which constitutes his physical freedom, and his liberation from the bondage of inborn instincts and desires, which constitutes his spiritual freedom, concomitantly make up what he altogether possesses of freedom. On the other hand, since Mankind is as a whole rooted in the same principle of body-liberation, it follows that every Man has his proper share in the human attribute of freedom, yet only to that extent to which he is capable of setting himself free from body-compulsion physically and spiritually. This applies no less to Man of today than to Man of prehistoric times. Even Primeval Man must have already enjoyed some limited measure of freedom when by the use of his crude tools he endeavoured to put the human principle into operation.

This deduction lends new weight to the old definition of Man as a "free being" in distinction from the "unfree" Animal. However, the concept of "freedom" is now understood in the new sense of "liberation from body-compulsion", as against its old interpretation in the sense of "liberation from Nature". That Man, for all his freedom, will for ever be subjected to the laws of Nature is self-evident for everyone who believes in the evolution of Man as an essentially *natural* process.

Although human essence finds its truest and immediate expression in the acts themselves of body-liberation, that is, in the manifestations of freedom physical and spiritual, and is indirectly expressed in the means through which body-liberation is implemented, that is, the artificial tools material and mental, the fact must well be kept in mind that Man, even if he is no longer of animal rank, is yet derived from the Animal and is still heavily afflicted with "animal heritage".

We saw that the human principle seized upon this very heritage from animal ancestry—the body and its function—in order to utilize it for establishing upon it the new scheme of body-liberation as its somatic basis. The body thus served the human principle in providing the material foundation for starting it into existence and keeping it in operation; furthermore, the body is tendering all the functions indispensable for the preservation and propagation of life, such as breathing, blood-circulation, metabolism, procreation, etc.—altogether vital vegetative functions which Man shares with the Animal.

Those who argue that for this common organization the human body is of animal rather than of human distinction disregard the essential point that the human principle has shaped the organs of the body so sovereignly and distinctively to its own ends that the form of the body has become a true reflection of this principle itself. The specificity of the human body, therefore, as reflecting the principle of body-liberation, is of a *fundamental* kind, just as is the specificity of the animal body which, on its part, is the true reflection of the animal principle of body-compulsion.

The fundamental diversity in the basic plans of the human and the animal body, in accordance with the fundamental diversity of their underlying principles, is not in any way qualified by the other fact that the organs of the human body, taken one by one, still reveal the same general pattern of structure and organization as the corresponding organs of the animal body. If we dismember the human body into its various parts, then, with the destruction of its ideal unity, nothing is left but the building material, the "animal heritage". Here we must beware of mistaking the somatic basis, which was used to build upon it a new whole, for the whole itself. A similar mistake would be if we smashed a marble sculpture to pieces and then argued that the statue was but marble and nothing else. Nor must we take the inherence of animal heritage in a sense, as if Man is only partly "Man" and still partly "Animal". As a marble statue is not partly marble and partly image, but throughout "image", so is Man throughout "Man". In the former case it is the artistic "idea", in the latter case it is the creative "principle" that moulds the raw material into a form of its own.

The shift from the animal to the human body-form will naturally be most conspicuous in the later stages of human evolution, when the specifically human form of *Homo sapiens* was reached;

but it must have started already when Primeval Man committed himself to the principle of body-liberation. Man did not gradually become "Man" by that slow process of evolution which converted his originally Apelike appearance into the modern human form, but he was already "Man" from the outset by virtue of his new scheme of evolution; and all that his further evolution was to do for him was not to make him more truly Man, but to make him more perfect as Man.

Just as Man is "Man" at any phase of his evolution and the Animal is throughout "Animal" at any phase of its evolution, it follows from the fundamental diversity of their evolutionary principles that their incarnations, Man and Animal, must likewise be fundamentally different from one another in any phase of their evolution.

This, then, is the final outcome of our inquiry: *Man is different in kind from the Animal; he constitutes a new order of life on earth, which, unlike the order of the "unfree" Animal, is rooted in "freedom".*

With this result, the old doctrine of the distinctive nature of Man has come to new life. But it is now on the firm ground of the facts of evolution that the uniqueness of Man has here been established, rather than on the old dualistic dogma which tended to split human nature into a human soul and an animal body on the presumption that the body was, so to speak, the cage in which the soul was chained. Attempts have been made to reconcile the old doctrine with modern evolutionary views in such a way as to hand over the human body, as a low product of animal origin, to the animal realm, if only the soul of Man, his true essence and distinction, was saved. One did not realize that the specificity of the human body was as distinctive of and as inseparable from Man as any of the distinctive attributes ascribed to the soul. Indeed, by surrendering the human body to the animal class one was, in effect, following a line similar to that of the staunch Darwinian writer Francesco Barrago (1869) who argued that "Man made in the image of God was also made in the image of the Ape".

Against any such misrepresentation of human evolution it

should be clearly understood that the physiognomy of the Ape, as the reflection of the animal evolutionary principle, is the perfect symbol of body-compulsion, whilst the physiognomy of Man, as the reflection of the human evolutionary principle, is the perfect symbol of body-liberation. In other words, the Ape is the image of body-compulsion while Man is the image of freedom—which makes all the difference.

Part Three

MAN MAKES HIS START

CHAPTER 14

The Mysterious Birth of Man

So FAR, the subject of our inquiry has been confined to the exposition of the exclusive principle that has been shown to be responsible for the singular heterogeneity of human evolution and, by its inherent contrast to the principle of animal evolution, to have made Man himself essentially different from the Animal. The question regarding the origin of Mankind has only been touched upon casually, although no doubt was left as to the author's strong belief in the descent of Man from animal stock.

If we want to continue with our inquiry and try to arrive at a rough idea of how the human principle might have come into existence and first development, we are at once faced with the tremendous problem as to the actual point of emergence of Man, a problem bristling with uncertainties and obscurities.

However, the start of Man is the most decisive event in human history. Naturally Man had first to spring into existence under the guidance of his selective principle before he was given the chance to advance on the line of this principle and so eventually to become what he is today.

Nevertheless, the start of Man is still the most mysterious event in human history. Although we may be able to form, by some lucky guess, a broad idea as to how Man, once the distinctive process of his evolution had been set rolling, would gradually have progressed in his specific development, we are yet entirely in the dark as to how this singular process came to be started.

We have no authentic information about the physical appearance of Primeval Man or about his mental condition; nor have we any direct knowledge about his immediate animal ancestry, or of

113

114   IN QUEST OF MAN

the place where his cradle stood, or the time when he was born—
so that everything that might be said about him can only be
purely a matter of deduction, if not mere guess or imagination.

In general, it is solely and only from our strong belief in the
soundness of the theory of evolution and in the overwhelming and
converging evidence regarding the genetic connections of Man in
relation to the animal world that we are persuaded to accept the
scientific postulate that at some time in the remote past Man
actually sprang from animal stock; then the privilege of being
Man's next of kin would fall to the Manlike Apes for their
demonstrable possession of the greatest number of fundamental
structural resemblances with Man.

Still, the lack of authentic material leads us to believe that
Primeval Man, the first representative of the human genus, must
for the time being be looked upon as an *hypothetic* figure; and
the same qualification applies to his not yet identified anthropoid
precursor.

Such being the problematical state of affairs in regard to the
momentous question of the actual emergence of Man, I am fully
aware of the grave danger incurred in trying to set out, if only in
theory, an imaginative picture of the special conditions and circumstances that urged the human principle, and along with it
Man himself, into existence. Any misinterpretation of facts or any
wrong deduction must lead to wrong conclusions and erroneous
conceptions; and it is because of this risk of drifting away from the
safe basis of proper research that we are reminded of the scientific
principle to build upon the "solid ground of well-authenticated
fact" rather than upon the "shifting sands of speculation". However, the trouble starts when, in the case of Primeval Man, we are
willingly but vainly searching for such well-authenticated facts.
From so remote a time, perhaps a million years ago, we cannot
expect to be presented with an abundance of relevant material. It
is true that thanks to the brilliant systematic work of palaeontologists over the last hundred years there has been assembled a small
yet highly impressive collection of fossil bones that seem to have a
more or less close connection with the early stages of Man.

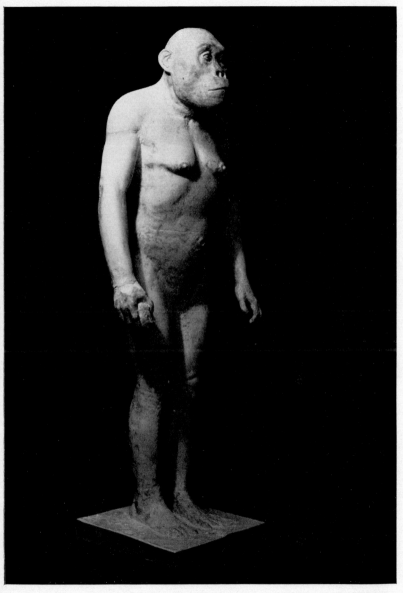

PLATE 1. Australopithecene. (By permission of the Trustees of the British Museum (Natural History).)

Unfortunately, most of the fossils are in a fragmentary condition and the difficulty arises how to sort out the various pieces of bone and determine their proper species. The impracticability of identifying early human bones by present scientific method will naturally become the more acute the nearer the bones are to the actual beginnings of Man, when most certainly they will still fall within the range of variation among the anthropoid Apes. So, if by some good luck we should actually happen to hit upon some genuine fossil remains of Primeval Man, we may not even be in a position to recognize their human origin, but would be inclined to attribute them to an Ape.

As an illustration of such a dilemma, I may refer to the sensational discovery made in 1924 at *Taung* in Bechuanaland, South Africa, of part of an infant skull which was outstanding for its

FIG. 7. A Taung skull. (From Ardrey, *African Genesis,* Collins.)

strange mixture of truly anthropoid and truly human features and yet was associated with a rather small brain cast. R. Dart, when investigating the skull, was of opinion that it was close to the human threshold, yet still the skull of an Ape, and accordingly labelled its genus as *Australopithecus Africanus*, the South African Ape. His classification was confirmed by the great authority in human pre-history, Sir Arthur Keith, who after carefully weighing each detail, taken one by one, also arrived at the final conclusion that the skull was after all that of an Ape. Later on, when many more fossil remains of the same, or closely related, species were discovered, it became certain without doubt that the Taung creatures were erect walking, and from heaps of cracked Baboon skulls it was inferred that they hunted and ate the Baboons. This story of Apes hunting and eating other Apes was readily taken up by an American writer, R. Ardrey, who quite logically and legitimately ended up his Book *African Genesis* (1961) with the absurd thesis that Man sprang from a predatory Ape from whom he inherited his killer instinct.

There is obviously an acute weakness in the scientific method of investigating such borderline cases on purely *morphological* lines,

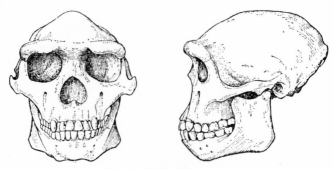

FIG. 8. Skull of Java Man.

(*From* Mankind in the Making *by William Howells, drawings by Janis Cirulis. Copyright © 1959, 1967 by William Howells. Reprinted by permission of William Howells, Doubleday & Co., Inc., and Martin Secker & Warburg, Ltd.*)

that is, by merely comparing and measuring, however carefully, their anatomical details. Even at a more advanced stage of human evolution it may still be difficult to reach a definite classification because of the inadequacy of the fossil material. As such an instance I may briefly refer to the equally sensational case of *Pithecanthropus erectus* whose skeletal remains discovered by E. Dubois in 1891 near *Trinil* in Java were first claimed to represent the much-sought "missing link" between Man and Ape, and hence was named "erect walking Ape-man". Other authors, however, would not agree with Dubois and rather insisted that the "Trinil Ape" by reason of the close resemblance of his skull cap with that of the Gibbon was a true Ape and not half-way on the human

side. It is characteristic of the inconclusiveness of purely morphological considerations that Dubois himself, in his later years,
reversed his "missing-link" theory in favour of the "Ape" theory,
and since the skull was much too large for that of the living
Gibbon, he put the skeleton down as that of an extinct *Giant
Gibbon*. From the point of view of comparative morphology no
objection could be raised to the liberty he took to simply multi-
plying the size of a bone for the sake of classification. From a
biological point of view, however, it would be as difficult to
imagine such a large Ape not having developed the huge bone
ridges on his skull that we find with the large Primates such as
the Baboon and the Gorilla as in the *Taungs* case it would be
difficult to imagine a large Ape hunting and eating other large
Apes without possessing the powerful teeth of the Leopard, the
known hunter of the Baboon.

From these criticisms it would appear that the *morphological*
method of comparison and measurement, although of recognized
and indeed indispensable value in cases of definite generic char-
acterization, must yet be completely inadequate and inconclusive
in those borderline cases in which the "diagnostic" features are
still indistinct, or missing. Here the *biological* method may prove
helpful; particularly as it is only from biological points of view
that a practical possibility is given to discern the first Man from
the last Ape, and so, at least in this respect, to lessen the uncertainty in the assessment of very early human bones. In our task of
investigating the emergence of Man the biological method will
therefore be widely applied, and so we may conveniently start
our investigation with the biological definition of *Primeval Man*.
According to our previous discussion, Primeval Man may be
defined as the *one descendant of an anthropoid stock who
integrated tool-use into his evolutionary mechanism*.

On first sight, one might perhaps be inclined to think that not
much is gained by so academic a differentiation of Primeval Man
from his animal ancestry. We shall, however, see that our definition admits of far-reaching implications. Not only do we possess
in it a practicable means for guidance in certain borderline cases

when an archaic skeletal fragment, which from morphological aspects suggests itself as of anthropoid origin, proves its anthropoid nature also from biological points of view, or otherwise may from these points of view suggest itself as possibly, or probably, or definitely of human origin. Our definition will also be helpful in dealing with the intricate problem of how it came to pass that from a common parentage the one descendant developed into Man and the others into the Manlike Apes. At the same time it will be useful to reduce to tolerable limits the vagueness and riskiness connected with any speculative reflection upon the mysterious birth of Man.

Still, I should like to emphasize the obvious point that any theory, such as the following, since it largely rests on deductions drawn from a choice of facts, must be taken merely in the sense of a working theory which serves to bring the various isolated facts into a logical and unifying relation one to the other, and thus to throw new light upon the ways and workings of Nature: yet which has to be modified, if not entirely discarded, when disproved by new discoveries. As it is, any fact, such as the find of a fossil fragment, gains significance only by its interpretation through deduction; and it is for this reason that science cannot dispense with theories and hypotheses. By the same token the following excursion into speculative fields may find its justification.

CHAPTER 15

The Hypothetical Common Ancestor

IF WE submit to the theory that Man sprang from animal stock, our next question will be, from which stock? Evolutionists are generally agreed that the closest relationship of Man is with the Mammalian Order of the Primates comprising the Lemurs, Monkeys, and Apes. Yet here their agreement ends. While most authors would grant the privilege of being Man's next-of-kin to the Manlike Apes, others insist that the structural resemblances of these Apes with Man may very well be a mere matter of parallel evolution, and that Man's true origin goes much further back, indeed should be sought for in an ancient Lemuroid stock of the *Oligocene,* or still earlier, in the primordial Mammals of the *Eocene,* that is to say, some twenty, or thirty, or more million years ago. Their argument, as stated by the noted American anthropologist H. Fairfield Osborn, runs like this: "An ancestral form must have a hundred per cent of the characteristics and potentialities of development which are observed in the descendant"; and from the fact that Man still shows in his structure some primitive features which are absent in the Manlike Apes and present in the lower Monkeys and Lemurs, the conclusion has been drawn that Man cannot possibly have shared the same ancestry with the living anthropoid Apes.

If we look upon evolution as a continual process of differential changes, in which old features are lost by the one offspring and retained by the other, and new features developed by the one and not developed by the other, we arrive at a different view; namely, that the fundamental structural resemblances between Man and the Manlike Apes are so overwhelmingly great and intimate that they cannot reasonably be explained otherwise than on the assumption of a close genetic relationship which would subsequently split up into differential courses of evolution. How this relationship may have been in detail is again an open question. Some authors are inclined to link Man genetically with the Gibbons, others with the Gorilla, Chimpanzee, or Orang. There

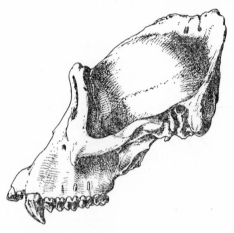

FIG. 9. Side view of skull of an adult male Gorilla.

is, however, general agreement on this point, that Man would not have evolved from any one of their present forms possessed of excessively long arms and hands and of powerful canine teeth. In our search for the *Common Ancestor* we should therefore go back to such a remote stage of evolution in which the specialization into the present specific forms of both the Manlike Apes and Man had not yet started.

The assumption is hereby made that the evolutionary process was not confined to the evolution of Man (of which we have a fairly good knowledge) but would equally have involved the evolution of the Manlike Apes (of which we know very little). To refrain, for lack of authentic fossil material, from speculating on the different ways of evolution taken by the Manlike Apes and to content ourselves with a mere description of their present forms—stating, for instance, that the Gibbon is a tiny and the Gorilla a powerful Ape and that both have great canines and excessively long arms in common—would not solve the actual problem.

However, so long as we know nothing definite about the immediate ancestry of Man, whether it was with this or with that form of the modern Apes, or with all of them, or with neither of them but with some extinct anthropoid stock, any statement about the common ancestral form must naturally depend upon deductions rightly or wrongly drawn. At the moment, therefore, it may seem advisable, if only provisionally for practical reasons, to hold on collectively to the whole group of the Manlike Apes, since they all are, if not the true descendants, yet at any rate close relatives, of the Common Ancestor. As this book is concerned with the history of the Apes only in so far as it affects the start of Man, a rough sketch outlining the presumable "characteristics and potentialities" of the common ancestral form should suffice to give an approximate idea of the differential trends of subsequent evolution leading to the present specific forms of both the Manlike Apes and Man.

Another (complementary) way of arriving at a well-defined though still highly speculative picture of the Common Ancestor would be by anatomical comparison of the skeletons of Man, recent and fossil, with those of the Manlike Apes.

From such a balanced comparison of human and anthropoid anatomy some authoritative evolutionists have indeed drawn a tentative picture of the *Common Ancestor* that showed him as a "primitive generalized Ape", not yet specifically specialized and hence still free for development into either Man or Ape. Thus J. H. McGregor visualized him as an Ape of fairly large size, with canines less specialized than in modern Apes, with arms shorter than in the modern forms, and legs shorter than in present Man, with hands of moderate proportions without the disharmony in the relative size of thumb and fingers, and with feet of the grasping type.

In order to ascertain the differential trends of evolution for both

Man and Manlike Apes we should not even confine our inquiries to the one group of the Manlike Apes. Since all Primates are more or less closely linked by phylogenetics, they all may afford valuable information. Thus we notice among them certain *general* features, such as the progressive development of the brain, the upright posture, the possession of hands and their manifold use— features, that is, which must have had significant bearing upon the making of Man. There are among them also certain *individualistic* features, such as the loss of the tail, the flattened chest, the lengthening of the forelimbs, the formation of huge bony ridges on the skull—features which are not uniformly shared by them, but are selectively distributed among them and thereby may link members of different groups closely together, while separating members of the same group widely apart. For instance, the lack of a tail separates the Barbary Ape from his true cousin, the Baboon, and links him with the genetically more remote Manlike Apes; or the development of huge bony crests on the skull roof separates the Gorilla from his true cousin, the Chimpanzee, and links him with the genetically more remote Baboon. From those individualistic tendencies we may well conclude that it was a *parallel* course of evolution that led to the development of identical features in separate groups. On the other hand, if within one and the same group some members developed such individualistic features and others not, we may from their absence conclude that those non-affected members would in their evolution have kept more closely to the original form from which the selective development started. Thus the round and crestless skull of the Chimpanzee points to an ancestral form of a similar smooth skull, whilst the Gorilla with his high crests (in conjunction with his huge canine teeth) would in this respect have specifically deviated from the original form.

In the following attempt to account, by way of deduction, for the biological facts responsible for the diversity of forms among the present Manlike Apes we may conveniently start with a brief survey of the anatomical and biological characteristics of these Apes.

The *Gibbons,* near 3 feet in height, are the smallest in the group, and of particular interest for their extreme degree of arboreal adaptation. Arms and hands are of such monstrous length that, when standing erect, the Gibbons touch the ground with their finger-tips. It is characteristic for such excessive degree of arboreal adaptation that the forearms have become even longer than the overlong upper arms. The hands, too, are extremely long, narrow and curved, and as they are being used like hooks for suspension the thumbs have become degenerate and in some species have altogether disappeared. The grotesque length of the forelimbs is in line with the peculiar method of climbing adopted by the Gibbons, particularly in escape from enemies. They only use their arms and hands when swinging from bough to bough, whilst keeping their legs firmly tucked up. This singular kind of *brachiation* enabled them to save their feet from high-grade arboreal modifications, which again accounts for their great ability to walk upon the ground in an erect manner with their long arms sideward outstretched as balancers.

Nearest to the Gibbons, in the high degree of arboreal adaptation, comes the *Orang.* When standing erect his finger-tips reach down to the ankles, and with him, too, the forearms may be longer than the long upper arms. His narrow and curved hands are also of great length, and here again with the exception of the thumb, which is "diminutive" and in its high position at the base of the long hand gives the impression of having "slipped up the arm". His feet take an equal part in his arboreal make-up. They are long and arched and turned inward to such an angle that the soles face each other. This makes walking upon the ground difficult, and it is only occasionally that the Orang descends from the trees and then, still in an erect posture, uses his long arms as crutches to swing his body forward between them.

With the *Chimpanzee,* arms and hands are again of great length; yet in an erect position his finger-tips frequently do not reach further down than the knees, and his forearms are, with rare exceptions, not longer, and may even be slightly shorter, than the upper arms. His arms are long and narrow with comparatively better thumbs, and his feet are only slightly inverted. Altogether, his climbing specialization appears to be less advanced than that of the Orang and he is able, if only for short distances, to walk erect with the flat soles on the ground.

The *Gorilla* has arms of such gigantic length and strength that on first sight they would seem to equal, if not to exceed, the overlong arms of the Orang. Looking, however, more closely into details, we find that his arboreal specialization is least advanced among the Manlike Apes. That in an erect posture his finger-tips hardly reach down to the knees may be due to the great length of his powerful torso. Still, his upper arms are longer than his forearms, his hands comparatively short with broad palms and fairly well-developed thumbs, and his feet are also of moderate length and less inverted. It is in keeping with his lesser grade of arboreal adaptation that the Gorilla is much given to terrestrial life. When walking upon the ground he supports his heavy body with his long arms, but may as well walk fully erect, particularly when going in to fight. He then, we are told, rises to his feet, and thumping with his fists upon his inflated chest and emitting frightful yells he rushes upon his enemy so as to lacerate him with his formidable teeth and at the same time strangle him with his powerful arms.

In this context it would seem that his gigantic arms, apart from their climbing properties, are an essential part of his fighting make-up, having grown so long and strong to match the colossal growth of his trunk which brings his total height, in an erect position, up to 7 and more feet. His special fighting method must call for steadiness in bipedal walk and a firm stance on the ground. Accordingly we note in his case a number of specific features which by their clear relation to upright gait must be taken as true *terrestrial* characteristics such as the small mastoid processes on the skull, the slightly convex curvature of the lumbar spine, the greater width of the pelvis and stronger development of buttocks and thighs, and the greater length and width of the heel region—altogether structural peculiarities that we find with no other Ape, and only, in a higher degree, with Man.

The skull of the male Gorilla is large and heavy, and is crowned by a huge median crest thrown up along the middle of the top and which is joined by another huge crest running across the back of the skull. Both crests serve to provide the necessary space for the attachment of powerful muscles of jaws and neck. The jaws, too, are large and heavy, and protrude in a muzzle-like fashion so

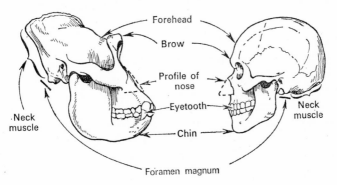

FIG. 10. Skull of a Gorilla and of Man.

(*From* Mankind in the Making *by William Howells, drawings by Janis Cirulis. Copyright © 1959, 1967 by William Howells. Reprinted by permission of William Howells, Doubleday & Co., Inc., and Martin Secker & Warburg, Ltd.*)

as to make room for a tremendous set of teeth with enormous tusk-like canines. As against this picture of terrifying strength, the skull of the infant Gorilla, with its humanlike set of milk teeth and its smooth and well-vaulted roof, resembles closely that of the human child; and also the female Gorilla, with teeth of comparatively moderate size and strength, keeps fairly close to the humanlike type of the infant skull, yet commonly shows a slight transversal elevation, and only rarely also a median ridge.

With the Orang, these divergent "sexual" characteristics are along similar lines, the infantile skull closely resembling the human form, and the female skull, with a very low transversal but no median ridge, remaining near to it; while the skull of the old male carries a huge median and transversal crest, and is also possessed of large and tusk-like canine teeth. However, whereas in the male Gorilla the fighting characteristics are growing fast during the comparatively short time of puberty, so as to reach their full power with maturity, this development continues with

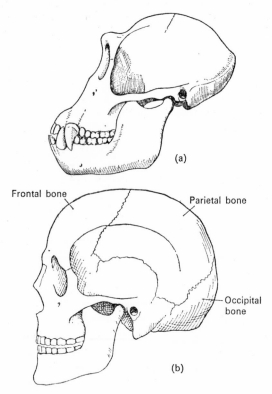

FIG. 11. Skull of a Chimpanzee (*a*) and of Man (*b*). (By permission of the Trustees of the British Museum (Natural History).)

the Orang throughout his life and may not even in his old age attain its definite completion, when the median crest may still consist of two separate, not yet united, elevations.

The Chimpanzee, measuring up to something like 4 feet, does not develop any such crests on his skull, nor are his jaws and canine teeth so much projecting. Indeed, both the male and the female skull keep in their general appearance very closely to their infantile form of great human-likeness, and it is only with old and strong males, reaching a height of up to 5 and more feet, that the canines may grow much larger, and a low transversal crest, very rarely also a slight median ridge, may develop on the skull.

If from the striking diversity of skull structure among the Manlike Apes we are to draw an explanatory conclusion, it is this, that with the Chimpanzee development has just arrived at a point marking the first formation of fighting characteristics—a course of evolution which is greatly advanced, but still in flux, with the Orang, and is fully completed with the Gorilla on a line similar to that taken by the Baboon and Mandrill. The question arises why the Gorilla and the Orang (and also Baboon and Mandrill) have acquired their formidable canine teeth whose growth caused the grotesque transformation of the whole skull. Evidently it was not for the sake of feeding, for the simple reason that all these Animals live preferably on a vegetarian diet. Darwin invoked his principle of sexual selection, the fight of the males for the females; yet he readily admitted that the teeth were also used as weapons in defence. Many observations, particularly on Baboons, have indeed shown that the teeth were actually employed for that purpose. Of the Gorilla it is said that at night he would sit on guard under the tree in which his family has its nest and keep off all dangerous beasts with his terrifying strength. Some Gorillas are even known to build their nests on the very ground under the protection of the males.

Our brief survey regarding the structural and behavioural peculiarities among the Manlike Apes allows making some certain deductions as to the original form from which their divergent evolution originated.

First, from the distinct arboreal specialization existing in the Manlike Apes and non-existent in other Primates, it can be deduced that the Common Ancestor lived in the forest, had the flattened chest of the brachiator, and arms somewhat longer than the legs; and from the Gorilla's least advanced aboreal development it can further be deduced that the normal proportion of longer upper arm and shorter forearm was still maintained together with a comparatively great width of hand and a fairly well-developed thumb; also that his feet were not yet greatly affected by arboreal adaptation.

Second, from the observation that the enormous growth of the canine teeth and the correlated profound change in the skull architecture are completed only in the Gorilla, not yet accomplished in the Orang, and only in their first beginnings in old male Chimpanzees, while still missing in the ordinary Chimpanzee, it can be deduced that such growth of the canines had not yet started with the Common Ancestor, who therefore had still the smooth and round skull and the moderate canines shown by the average Chimpanzee. Moreover, since the enlargement of the canines, as a token of fighting spirit, appears to be dependent on a certain measure of physical size, bulk, and strength like that shown by old male Chimpanzees, the further deduction can be made that the Common Ancestor was of somewhat smaller size than those fighting Apes, that is, approximately of the size of the average adult Chimpanzee.

Third, from the observation that all his offspring show the disposition to upright posture and ability of bipedal gait it can be deduced that the Common Ancestor likewise possessed this disposition and ability, thus allowing his progeny, if the occasion arose, to integrate bipedal locomotion into their evolutionary mechanism—which so happened to be the case, in a slight degree, with the Gorilla and, in a larger degree, with Man.

Fourth, the observation that disposition to upright posture has enabled all higher Primates to use their hands freely, in particular for picking food and also for grasping and throwing stones and other objects in play and attack, leads to the deduction that the Common Ancestor would have likewise used his hands for any such purpose, and thus allowed his progeny, if the occasion arose, to integrate extra-bodily defence into their evolutionary mechanism—which so happened to be the case with Man.

In short, the Common Ancestor, hypothetical as so far he is, presents himself in our theory as a "primitive Ape" not so much dissimilar to his picture drawn by McGregor, so that even those who hesitate to agree to our theory may still agree to that picture. It should, however, be mentioned that our summary exposition of the Manlike Apes rather suffers from certain oversimplification. One point is that there exists such great variability among the Apes, not only in size and body proportions, in hair and colour, but also in habits and behaviour, that when speaking of the Gibbon, the Chimpanzee, etc., we can only refer to a typical form of these Apes.

Another point is that the Gibbons are possessed, on the one hand, of a smooth skull without any crest formation, and on the other hand, of very large sabre-like canine teeth which do not readily fit into the picture of these tiny Animals and their accomplished virtuosity in escaping from enemies by their exceptional climbing technique. How and why the Gibbons acquired their long canines we do not know. Perhaps they were derived, as it has been suggested, from some larger and stronger ancestral form with aggressive tendencies.

The Gibbons differ also in other respects from the pattern of the Manlike Apes. They appear to be of a more primitive type, and this applies particularly to their brains. It is for this reason that they are frequently classified apart from the others in a separate group, in which case they would no longer be in question for an immediate common ancestry with Man.

Also the Orang, although a legitimate member of the anthropoid family, is frequently assumed to have at an earlier time departed from the common ancestral stock. This assumption would be consistent with a double-phased evolution of his, in developing first his excessive arboreal equipment and later, when growing larger and stronger, adding his great canines. His early segregation leaves the two African Gorilla and Chimpanzee as the only claimants for a direct common ancestry with Man, which, in the circumstances, would help to simplify the problem of his emergence from Apehood.

CHAPTER 16

The Start of Man

To A "primitive generalized Ape", such as the Common Ancestor is outlined in our theory, several ways were open for further development. If he were of small stature and sought safety by escape he could have taken to his climbing propensities and developed a high-grade pattern of arboreal adaptation, like the Chimpanzee; or when later on he increased in size and bulk and felt strong enough to show fight, he could have added to his arboreal make-up a set of powerful teeth, like the Orang. On the other hand, if from the outset he was of larger and stronger build, he could immediately have developed his fighting potentialities and become a huge and powerful Ape with enormous canine teeth, like the Gorilla; or he could have taken up active defence, not by means of his own body, but by means taken from outside, as when he would beat off a dangerous predator by hurling a shower of missiles upon it. *This, I propose, was the way taken by Man.*

What was actually the situation from which Man started off upon his new career, and how he could manage to survive the struggle for existence with his improvised weapons, are questions difficult to answer for lack of direct evidence.

It has been argued that Man would never have become "Man" in a forest-environment. As H. Klaatsch once said, "In the trees Man would inevitably have become an Ape." If we put this phrase the other way round and say that in the trees the Ape would never have become Man, the situation is to my mind well grasped, if only for the reason that fighting with extra-bodily means had, at least initially, the constant availability of those weapons as its prerequisite, and it was only on the rocky ground that stones were lying about in abundance and were ready to hand.

However, habitat on the ground does not in itself solve our problem. Baboons, we know, are perfectly terrestrial Animals without, however, having ever made an approach to "hominization".

Some authors, therefore, have brought the rise of Man in causal connection with his *descent from the trees*. Their theory is that it was some kind of cataclysmal event with drastic change of climate that led to the dwindling of the tropical forests, as a consequence of which Man found himself suddenly forced to the ground where he had to face a totally different milieu heavily loaded with grave menaces to his life and with new needs to overcome.

In such a perilous situation, deprived from the protection of the dense forests, he would have lived in social groups like many Primates, and when attacked by a dangerous predator would have acted in concert with his co-fellows and in a supreme common effort hurled any stone he could get hold of upon the attacker. The superiority of the artificial tool over the bodily organs came to his aid in allowing him to deal with the attacker *from a distance*.

Again, habitual practice would lead to improvement, not only in skill, but also in technique. At first, emerging Man may have picked up the stones just when he needed them and where he found them. In time he may have learned by experience to select the more suitable stones and may even have collected them in advance and carried them with him on his wanderings. Such anticipation of later need would not be exceptional. Brehm, for instance, reported of Baboons which, enraged by rifle-shots, picked up every stone on their way and rolled them down upon their attackers. An old male was actually seen climbing a tree with a large stone under his arm and then hurling it down.

Experience may also have taught him that stones, splintered accidentally, were particularly effective, and this knowledge may have induced him to break stones deliberately. Apart from stone, wood may have played an important part in defence. Also the

F

long bones and horns of Animals may readily have served as early weapons, a suggestion made by Dart for the *Australopithecinae*.

Leaving aside any further guesswork and taking a long view of the early stages of human evolution, we may broadly say in the words of W. D. Wallis that "Man began using wood and stone and then learned to improve upon natural tools by slightly modifying them. Improvement led to manufacture, and the tool-user became a tool-maker."

In principle, the start of Man appears to have happened simply by his adhering tenaciously to a method of defence that, as such, was well known among Monkeys and Apes, and occasionally used by them. But by making extra-bodily defence his exclusive life-protecting device, one that was instinctively adopted, practised and furthered by each following generation, he created a new situation in which the use of artificial tools *became automatically incorporated in his evolutionary scheme.*

With the subsequent decline of climbing dexterity, Man became inevitably and irrevocably chained to the ground where he had to stand or to fall with his artificial tools. Thus he found himself caught in the meshes and the iron grip of an evolutionary principle that we now recognize was the principle of body-liberation, and as such conferred human rank upon him. Unwittingly and automatically he divorced himself from his former animal kinship, when, with his raw weapons in his hands, he clumsily shuffled along the thorny path that was to lead posterity to power and greatness.

Seen from a higher perspective the birth of Man suggests itself as a most natural and straightforward event. In a sense his evolution ran parallel to that of the Gorilla. Both, Man and Gorilla, took primarily to fighting for defence, the latter resorting to his bodily means of growing powerful teeth and arms; Man relying on stones and wood and other improvised weapons which Nature offered him as tools equally efficient for warding off attackers. Both must have been possessed of such a great measure of physical size and bulk as to make them feel strong enough to venture upon active defence; and both employed arms and hands in their respective fighting methods and subsequently developed such specific characters in their bodily make-up as were obviously in relation to, and for the improvement upon, bipedal gait. All this makes the Gorilla, as it were, the "opposite number" of Man in the animal realm.

This parallel trend of evolution must not, however, be taken to signify a closer genetic relationship between the two. It may well be so, but evidence is lacking. It does, however, pose the question as to how it came to pass that the one took to his bodily means and the other to extra-bodily means for the same purpose of active defence; or to put it another way, why the Gorilla, and for that matter any living or extinct Ape, failed to steer into the human track of evolution.

One thing seems clear that it was not within their own dis- cretion to choose between the two alternatives. Rather are we led to believe that their option of different fighting schemes was a compulsory and indeed desperate act in the struggle for existence forced upon them by different environmental circumstances. For Man, we may assume, it was his being cut off from the protection of the trees that urged him into the channels of extra-bodily defence; for the Gorilla it was presumably his continued habitat in the forests that kept him within the grip of his own bodily powers. Our theory thus implies that Man and Gorilla have started upon their respective careers, not the one under the eyes of the other, but in different places and under different conditions and circumstances. Either the region in which the Gorilla lived was not affected by the decline of the trees, or he happened to have found his way back to non-affected forests, while Man was inescapably tied to the ground.

As with growing specialization in tool-use and upright gait, Man was "stuck" in his human career, so was the Gorilla, once his canine teeth and arms were growing to powerful strength, "stuck" in his animal career. The suggestion has been advanced that Man may have derived from an Ape with large tusky canines which were gradually reduced in length and strength through the increased use of artificial tools. This suggestion even goes back to

Darwin himself when he argued that "the early progenitors of man have been once covered with hair, both sexes having beards; their ears were probably pointed, and capable of movement; and their bodies provided with a tail. . . . The males had great canine teeth which served them as formidable weapons."

From his reference to the presence of a tail it would seem that Darwin here had not the immediate ancestry of Man in mind. Also it may be objected that his suggestions are nowadays rather antiquated. Still, the most essential feature of his picture, the great canines possessed by the males as formidable weapons, is far from being obsolete today, since *Dryopithecus*, the *miocene* Ape with apparently tusk-like canine teeth, is still widely believed to represent the ancient anthropoid stock from which Man sprang.

In the light of the theory advanced in this book no such large canines have ever been in the direct ancestorship of Man. Their powerful growth must be taken as a token of well-advanced, hence irreversible, evolution: the outward expression of an innate indomitable urge that will never cease to keep the Animal under its sway, wherefore the Gorilla has for ever missed the chance of becoming Man.

The uncertainty with regard to the nearest animal relationship of Man has encouraged some authors to play with the idea of a multiple origin of Mankind. Thus the yellow-reddish Asiatic Orang has been linked up with the Malayan, and the two blackish African Gorilla and Chimpanzee with the tall and small Negroes respectively. Another theory tried to connect the massive *Neanderthal Man* with the Gorilla, and the slender *Aurignacian Man* with the Chimpanzee. Also the multitude of human races was taken to refer to a multiple origin of Man which was supposed to have occurred at different times and in different parts of the earth. All those theories are based, more or less, on unduly superficial comparison. Against them stands the fundamental uniformity in anatomical structure among the present human races, a criterion strongly suggestive of a monophyletic origin of Mankind. However, since next to nothing is known about the immediate precursorship, the theoretical possibility of a multiple origin cannot at present be repudiated.

So is the question still open regarding the birthplace and the birthdate of Man. When early human bones were found in Java and China, dating back to mid-*pleistocene* times, perhaps 500,000 years ago, opinion was in favour of Asia; but since the discovery of *Australopithecus*, living at a still earlier time, Africa was hailed by many authors as the presumable cradle of Mankind.

There has also been a lively debate over the question regarding the particular organ which would have given the actual impetus to the birth of Man. Most popular is the view that it was the *brain*. Evidence indeed goes to show that in the long stretch of Primate evolution it was conspicuously the brain that attained to an astounding degree of expansion and elaboration, particularly with the Manlike Apes in whom it reached a high state of organization, something like a smaller replica of the human brain. If the argument had been that it was from the high organization of the anthropoid brain that Man set out on his road, all would be well. However, the argument runs that it was the *large* brain, that is, a brain transcending in size and structure that of the Ape, which launched Man into existence. Man has become "Man", we are told, because he was an "Ape with an overgrown brain" and with a corresponding preponderance of higher mental qualities.

How the "big brain" could possibly have come about was difficult to explain. Darwin already struggled with this problem when he argued that "it might have been an immense advantage to man to have sprung from some comparatively weak creature", since "an animal possessing great size, strength, and ferocity, and which, like the gorilla, could defend itself from all enemies, would not perhaps become social; and this would most effectually have checked the acquirement of the higher mental qualities . . .".

On the similar ground of social life in connection with the large brain stand those theories which claim that Man happened to live, at first, in a region free from dangerous beasts where he was able to lead a leisurely social life and by unhampered mutual communication of experience and thought to develop his mental powers to such a height as to enable him when need arose to *invent* his artificial tools.

The criticism here is that the gregarious Monkeys, in spite of social life and mutual communication, did not develop their brains on any larger scale than on that in which we actually find them in their present forms. Other theories take rightly, and in concurrence with our own theory, the opposite view, namely, that it was hardship due to the stresses of environmental conditions that forced upon Man the urgent need of exerting all his wits if he was to survive, and that it was through those supreme mental efforts that the brain was stirred to fresh growth.

Here in any case due account is taken of the intimate interplay between tool-use and brain development in a situation which must have taxed the mental faculties to the extreme. Although we may reasonably assume that brain growth, this outstanding feature of human evolution, had already occurred with the start of Man, it would not carry the implication that Man started from an already advanced brain condition, that is, from a brain that transcended greatly that of the intelligent Chimpanzee.

However, considering the great variety among the Manlike Apes in alertness and ingenuity, and also considering the dominant role that the brain must have played from the very start of early Man, we might well assume that the "brain led the way", for without a certain high degree of intelligence Primeval Man would not have managed to be successful in his new way of life.

In conclusion, Man may have started with a certain higher grade of intelligence, but contrary to any such idea that he launched upon his evolutionary career by his own will, reflection, or invention, owing to an enlarged brain, the author of this book takes the view that Man slipped into his specific scheme of extra-organismal defence as blindly and automatically as did the Gorilla into his opposite scheme of organismal defence.

Another organ made responsible for the emergence of Man is the *foot*. Here the argument is that his adoption of upright gait released his hands from their former duties of locomotion and made them free for unrestrained employment in the manipulation of tools and weapons. Thus, by "carrying his head high", Man differentiated himself from the Animal, which holds its head poised downward towards the ground. However, upright posture, in association with bipedal gait and freedom of hands, is not an exclusive characteristic of Man. The Gibbons are well able to walk fully erect, and the Gorilla even shows a tendency towards reinforcing his ability of bipedal action without developing into Man.

The *hands* have never been particularly suspected of being primarily responsible for the rise of Man, for the obvious reason that they are an ostentatious possession common to all Primates. Yet, if we were pressed to make a single organ responsible for the birth of Man, we should have to grant this privilege to the hand as the outstanding organ that with its grasping fingers and opposable thumb is virtually cut out for the wielding of tools and weapons, and indeed is indispensable to the scheme of tool-use. There is also good reason to believe that it was the possession of hands that uniformly prompted the Primates into their habit of erect posture, and along with stereoscopic vision, stimulated the progressive development of their brains. Still, as a feature peculiar to all Monkeys and Apes, the hands cannot be held responsible either for the emergence of Man.

From the standpoint of this book it would appear that the question as to which particular organ actuated the birth of Man cannot be answered correctly because it is wrongly put. There is no one organ by itself that would have brought Man into existence, although the brain and a certain height of intelligence, the feet and the ability of erect gait, and the hands with opposable thumbs had each of them an essential and integral share in his rise by establishing the prerequisite somatic and functional basis for the human principle from which to start and to continue. The proper question would be: Which was the primary factor responsible for the emergence of Man? And here the answer is that it was the shift from the animal to the human principle of evolution. For it was in pursuance of this principle that Man divorced himself from the Animal and thus created a realm of his own.

If this drastic shift from the animal to the human principle is rightly understood as signifying the start of Man, we will realize that there was never such a thing as an "Ape-man", if by this concept is meant a creature structurally and culturally intermediate between Ape and Man. H. F. Osborn, some years ago, caused a sensation when he came out with the thesis that Man has always, from his very start, been throughout "Man", and that the Ape-man was "but a myth". His challenge, however, fell flat because of his obstinate denial of any genetic relationship between Man and Ape beyond a very remote common root in some archaic Lemuroid stock. Yet his rejection of the intermediate Ape-man finds full support in our own theory. For, from the critical point in time, when Man consistently and persistently turned to the use of artificial tools in preference to, and actual abandonment of, the natural means of his body, and by so doing made tool-use the soul and destiny of his evolution, he was truly Man. Neither was he, in essence, pre-human or sub-human or something of that kind, although structurally he still closely resembled his anthropoid precursor, and culturally stood on the lowest human level imaginable. Nor was his anthropoid predecessor, in essence, either pre-human or sub-human or something like that, but he was still being held firmly in the grip of the animal principle of body-compulsion and was therefore a true Ape. It was manifestly the direct antagonism between their respective principles of evolution rather than the admittance of any intermediate stage which drew at once the radical line of demarcation between them.

Disregarding all detail and keeping only to the whole and the essential, the incident of the emergence of Man impresses us by its simplicity and its mechanistic working. The door to Mankind was wide open. The Ape was, so to speak, ready for Man. Man had indeed nothing more to do than to hang on tenaciously to a kind of defence known to him from other Primates and from his own casual experience. Nature did all the rest for him when she perpetuated his new scheme through her selectively sifting and intensifying agency. Thus Man started into existence as a fighter, and how hard had he to struggle for his survival with his poor weapons. The widespread notion that he was born a hunter is not supported in this book. Rather was he himself the hunted prey, and it was only after a time, with improved technique in tool-use, that he was able to reverse the situation and, in search of additional food, make animal-hunting a favourite pursuit.

CHAPTER 17

Theory and Fact

WE NOW pass from theory to fact, and deal with the direct evidence relating to the early development of Man and his anthropoid ancestry such as is tendered by a still scanty but steadily increasing array of skeletal remains believed to be in more or less close phylogenetic connection with human pre-history. This chapter, which is pre-eminently concerned with the question as to what help our theory might offer for the identification of archaic fossils, will only state issues of fundamental relevance. The whole story of human pre-history as yet known may easily be looked up in special textbooks on prehistoric Man.

From a time so remote as a million or so years ago, when Man is believed to have entered the scene, we cannot expect to find complete skeletons that would give a true and easily recognizable picture of early Man. Rather must we rely on some fragmentary pieces of skull, jaw, or limb, or even on a single tooth, and try to fit them in appropriately. Their identification meets with the further difficulty that it may be hard, if impossible, to differentiate the remains of early Man from those of his anthropoid predecessor, since anatomically they would at first have been very much alike. Only after a long period would the first recognizable human features have emerged. Even so, those "diagnostic" features, truly human as they were, would in their initial stages still fall within the range of variation of the Anthropoids, which again must hamper their proper interpretation.

The finding of artificial tools, in association with skeletal remains, may help identify early human bones. But here the difficulty arises not only that worked stones are not always found in association with the bones, but that the stones may still be lacking the unmistakable signs of deliberate flaking and hence are not yet identifiable as Man-made. Again, it may have taken Man a considerable length of time to reach the advanced stage even of a crude, yet recognizable, chipping of the stones.

Not so long ago there was a debate over the question of whether Apes, too, might have been responsible for primitive stone-flaking. From the standpoint of this book any such suggestion is firmly rejected. The manufacturing of tools, however crude in the beginning, is not a matter of casual use of tools, but is the logical result of a very long period of development in which the process of evolution was intensely and consistently focused upon the employment and perfection of artificial tools, and this was coupled with the adoption and perfection of upright gait and with the elaboration of the brain. It was from this distinctive trend of evolution towards tool-use that the impetus stemmed of improving the tools by eventually shaping them systematically. If, therefore, there has ever been an extinct creature chipping stones deliberately, it would have come under the human principle of evolution, that is, on the human side.

Quite recently a number of scientists, on the authority of K. Oakley, have come forth with similar views when they accepted tool-making as a definite criterion for humanity. Their argument is that systematic tool-*making*, as against simple tool-*using*, implies a stage of higher mental abilities, and for this reason bears out strong evidence of the existence of Man who made the tools. Still, even before Man came to fashion his stones, when he was only a tool-*user*, he was already truly Man working incessantly, within his evolutionary trend, towards the perfectioning of his scheme of tool-use.

If no worked stones are found associated with ancient bones and there is doubt as to whether the bones are of the human or the animal order, which clues does our theory offer to help identification?

First, size is of importance. Active defence with extra-bodily weapons must, in its initial stages, require a good measure of physical size and bulk, such perhaps as is shown by old male Chimpanzees taking to active defence by their teeth. We should therefore expect early Man to have been of somewhat similar size

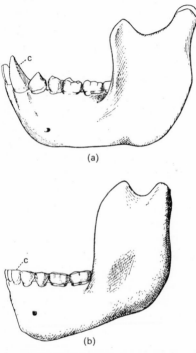

(a)

(b)

FIG. 12. The lower jaw of an anthropoid Ape (*a*) and one of the *Australopithecinae* (*b*). Note that the Australopithecine canine (*c*) is very different from that of an Ape in its shape as well as its size, and at an early stage of attrition has become worn down flat from the tip so as to be level with the other teeth (as in Man). (By permission of the Trustees of the British Museum (Natural History).)

and strength. It means that a fossil skeleton indicative of general body-size considerably smaller than that of the Chimpanzee would not probably be of early human origin.

Second, such a comparatively large body-size raises the question of the long canine teeth. Here, according to our theory, the development of these teeth to anything like the length and strength shown in the Gorilla, Orang, or Mandrill is contrary to the scheme of human evolution, and hence must be excluded from early Man, as well as from Man's immediate ancestorship. The same consideration applies to the accompanying growth of huge median and transversal skull crests of the Gorilla-type and which equally excludes Man. However, since in modern Man the canines are still slightly projecting, and also by virtue of their long and strong roots testify to a former stronger development, we must be prepared to find these teeth and jaws of some greater length and massivity in early Man than in modern Man. "Large canines" and an "Animal-like snout", both in comparison with the reduced sizes of teeth and jaws in modern Man, would therefore not in themselves disprove the human nature of a fossil skull provided their measurements do not exceed those shown by the Chimpanzee.*

Third, as regards the mode of locomotion we may have to reckon with some slight degree of arboreal adaptation. However, anything like the advanced arboreal adaptation after the pattern of the Gibbon, Orang and Chimpanzee would not fit into the picture of early Man or that of his predecessor. Rather would we expect the skeleton, if it were on the human side, to display some initial characteristics of terrestrial habitat in the direction of upright gait, not only on legs and feet but also on the skull, where the erect poise may already have led to a forward shift of the *foramen magnum* and the occipital condyles.

Fourth, although human evolution made evidently for brain expansion it started out from the animal brain. Therefore a comparatively small skull cavity and a flat vault would not yet contradict the human origin of a fossil skull, if this be suggested by other evidence; whilst a brain, greatly surpassing the anthropoid range, should be strongly indicative of its human nature even if the other evidence be inconclusive.

In general, comparatively large body size, relative smallness of

* In the genuinely human skull of *Java Man* we find a slight median elevation and in some *Paranthropus* specimens even a low median ridge, both being tokens of a heavy masticulatory apparatus, yet with low canine teeth and very different from the huge crests developed by the Gorilla.

canine teeth and jaws, a skull roof without huge crests, and the initial signs of erect gait and increasing brain volume are characteristic features which make up the specific physiognomy of early Man. Still, from the biological point of view, each such feature can *by itself* not yet be considered as conclusive evidence.

For instance, the fragment of a skull large enough to be in question for early Man and with no huge crests and no powerful canines might still be that of a female Ape. In this case, however, the brain capacity would be suspiciously small, too small in any case to be indicative of Man. Or if such a large skull exhibits feeble features of erect gait, it still could be the skull of an Ape, say, a male Gorilla who shows in his skeleton distinct trends to erect gait. In this case the canines would be suspiciously large, too large to be in question for Man. It is different if a large skull shows not only comparatively small canine teeth, but, in addition, also signs of permanent upright gait, such as a marked forward position of the *foramen magnum: this combination makes it absolutely certain that the skull was human.* Bipedal erect gait, if pursued to such an extent as to impress itself upon the skeleton, can only be tolerated by Nature if Man, or Ape, had such powerful weapons at their disposal as were necessary to hold their ground in the struggle for existence. It was only Man who saved his body from the development of any such powerful fighting organs by taking his weapons from without his body. In such case the brain would most likely show some initial enlargement, and this again would help to confirm the human origin of the skull.

In search for the possible precursor of Man the already mentioned genus *Dryopithecus*, a large Ape about the size of the Chimpanzee, living in the late-*miocene* and early-*pliocene* age, that is, about 25 to 15 million years ago, is still in great favour. When, in the middle of the last century, two broken lower jaw bones and part of an upper arm bone of this Ape were discovered in France, the conspicuous resemblance of the molar teeth with those of both Man and living Anthropoids, in conjunction with the moderate length of the arm bone, gave rise to vague speculations about his close relation to human and anthropoid phylogeny.

He was even hailed to be the hypothetical ancestor of Man and Apes. Judging, however, from the enormous remainder in one of the lower jaws, the canine teeth must have been of considerable size. K. A. Zittel described the mandibular symphysis as high, forward pointing, and narrow, "indicative of a considerable snout like that of the Baboon". From this evidence it follows that *Dryopithecus*, however closely he may have been related to the

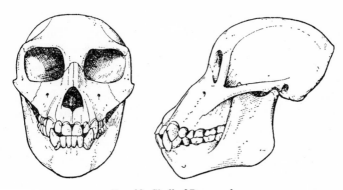

FIG. 13. Skull of Proconsul.

(From Mankind in the Making *by William Howells, drawings by Janis Cirulis. Copyright © 1959, 1967 by William Howells. Reprinted by permission of William Howells, Doubleday & Co., Inc., and Martin Secker & Warburg Ltd.)*

anthropoid predecessor of Man, certainly was not identical with him.

In late-miocene and pliocene deposits in the Himalayas, India, a number of anthropoid fossils have been discovered, mostly consisting of fragments of jaws and of teeth, some of which are described as very humanlike. Thus, E. G. Lewis referred to one of the jaws, labelled *Ramapithecus*, as approaching closely to the human type in the small size of the canine tooth and in the shape of the dental arch. This Ape, "almost at the human threshold", might perhaps have a chance to stand for the ancestorship of Man, if only we knew more about him.

More promising seem to be the numerous finds of early-miocene Apes, represented mostly by teeth and broken jaws, but also by fragments of other bones, that have been made more recently in Kenya, Africa. Some of the smaller types, such as *Limnopithecus*, show some resemblance to the modern Gibbons yet without their overlong arms and their brachiating characters already developed. A larger type, named *Proconsul*, of which also the greater part of the skull was preserved, an Ape with fairly strong canines and probably possessed of some ability towards upright gait, shows resemblance to the modern Chimpanzee and was claimed by his discoverer L. Leakey as the probable ancestral form of both the modern Apes and Man. On this assumption, we would with regard to Man have to leap over the enormous span of around 25 million years to arrive from *Proconsul* to the *Australopithecinae*.

In pliocene deposits of Italy, dating back perhaps 10 million years, the skeleton of an Ape of near-Chimpanzee size, *Oreopithecus*, was found which was described as a mixture of anthropoid and human features and by some authors even believed to stand on the human side. However, the circumstance that the skeleton was situated in a brown coal pit suggests strongly that this Ape had his habitat in the trees of swampy forest, which must exclude him from the human line. The possibility can certainly not be denied that the modern Apes, all or part of them, are the continuation of those miocene African Apes which were primitive in their organization and had still to develop their excessive arboreal specializations. There is, however, the risk of oversimplification. *Limnopithecus*, for instance, stood by his primitive limb proportions in line with the *Cercopithecinae*. But so did pliocene *Pliopithecus* 10 million years later, who also was claimed to be ancestral to the living Gibbons and yet was still as remote from them as his miocene form, *Limnopithecus*.

The *Pliocene* was followed up by the *Pleistocene*, also known as the *Great Ice Age*, which began around a million years ago. It is notably the Age of Man himself, and a good number of relevant fossils have come into our possession from this latest prehistoric period.

One of the first to be discovered, in the middle of last century, was *Neanderthal Man*, the "Man of the Old Stone Age", whose

PLATE 2. Gibbon. (By permission of the Trustees of the British Museum (Natural History).)

PLATE 3. Orang. (By permission of the Trustees of the British Museum (Natural History).)

PLATE 4. Chimpanzee. (By permission of the Trustees of the British Museum (Natural History).)

PLATE 5. Gorilla. (By permission of the Trustees of the British Museum (Natural History).)

bones were found in the Rhineland, Germany. There was, it is true, some doubt at first about his great antiquity. But in spite of his strange Ape-like physiognomy, which at that time was conceived as being due to pathological abnormalities such as rickets, idiocy, and what else, he has never been earnestly challenged as to his genuine hominid status.

It was not so with two other equally sensational discoveries of fossil remains dating from an earlier part of the Pleistocene: *Pithecanthropus erectus* living in roughly the middle of this

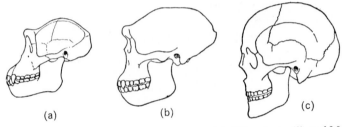

(a)   (b)   (c)

Fig. 14. Side view of the skull of a Chimpanzee (*a*), *Pithcanthropus* (*b*), and Man (*c*). (By permission of the Trustees of the British Museum (Natural History).)

period, about 500,000 years ago, and *Australopithecus Africanus* living at a still earlier period, perhaps 750,000 years ago. Both are so-called "borderline cases" to which, as we shall see, the biological method may be successfully applied. *Pithecanthropus* is now generally acknowledged as a member of the human species, hence a brief review of his case seems sufficient; while *Australopithecus*, still in the midst of controversy, needs to be dealt with in some greater detail.

As has already been mentioned, *Pithecanthropus* was discovered in 1891 in Trinil, Java, by E. Dubois. Only the cap of a skull and, at a distance, a thigh bone and some molar teeth were originally found. The teeth and the thigh bone were remarkably humanlike, but the skull with its flat and narrow frontal area and a brain capacity estimated at only about 800 c.c., or slightly more, was by no means of modern human type. Moreover, over the eye-sockets were thick brow-ridges bulging out, as in the Apes, and there was also a slight median elevation on the top of the skull. Still more significant, the back of the skull bent sharply downward and forward, thus giving the occipital bone an angulated shape with the foramen magnum well in a forward position. From this peculiarity it was rightly inferred that the owner of the skull belonged to an upright-walking species, which was also confirmed by the humanlike thigh bone. No artificial implements were found with the skeletal remains.

Was *Pithecanthropus* an Ape, or was he Man, or was he an intermediate stage between them as suggested by Dubois? The majority of scientists showed themselves at that time in favour of a transitional form, but quite a few were definitely for an Ape, and only very few were hesitatingly for true Man. As an Ape, he would have had double the brain capacity of a Chimpanzee, and also largely exceeded that of the male Gorilla; as Man, his brain capacity would only have been about half the size of that of modern Man.

If we apply the biological method we first note that the skull belonged to a strongly built elderly individual over 5 feet high. If he were an Ape of that size and strength, we would expect him to have developed fighting characters on the skull, which evidently he had not. Although the muscles of jaw and neck must have greatly exceeded those met with in modern Man, their attachment did not require the huge crests that are so prominent a feature of Gorilla and Orang. In consequence, the skull cannot have possessed the large and powerful canine teeth of these Apes. Next, there is the sharp bend at the back of the skull, which in conjunction with the forward position of the foramen magnum carries forcible evidence of upright gait.

Here then we have the combination of the three essentially human features—appropriate height and strength, absence of powerful canine teeth, and upright walk—which leads to the irrefutable conclusion that the skull was that of a *human* being.

Already in the first edition of my German book, 1922, and in a later paper,* when the controversy over the *"Trinil Ape"* was still in full swing, I made, on the above-mentioned biological grounds, a strong plea for the human origin of the bones. However, it was the smallness of the *Trinil* brain that discouraged most scientists from acknowledging the brain as a human brain. And yet it should have been a logical conclusion that in the beginnings of human evolution the brain development must equally have been in its infancy; moreover, the *Trinil* brain showed already an enlargement ostentatiously above the range of the anthropoid brain.

The climate of opinion only changed when in 1927 and following years *Sinanthropus*, the *Peking Man*, was discovered who was apparently a close relative of *Pithecanthropus* and living at almost the same, or somewhat later, geological Age. There was definite proof that *Sinanthropus* fashioned his tools and also knew of the use of fire; besides, his brain volume was in the range of 900 c.c. to 1200 c.c. From this evidence of advancing development in the human direction the conclusion was firm that *Sinanthropus* stood definitely on the human side. What was true of him should also be true of his cousin *Pithecanthropus*, now commonly called *Java Man*. Whether or not the finding of charred wood on the *Trinil* site is accepted as sufficient evidence of fire-use, it seems in any case reasonable to assume that *Java Man*, like *Peking Man*, knew of the use of fire and also made tools.

Of *Australopithecus*, discovered, as mentioned before, in 1924 at Taung in Bechuanaland, South Africa, there was found originally only the facial region and the adjoining base of an infantile skull, with the milk dentition still in place and the first permanent molars just cutting through, and, separately, the natural limestone cast of the brain. The skull was remarkable for its strange combination of truly simian and truly human features, and the question was, in which direction, the anthropoid or the human, the *Taung* child would have developed if he, or she, had grown up. R. Dart who first examined the fossil was of opinion that the skull, although well advanced in the human direction and probably

* *Pithecanthropus erectus—Homo trinillis. Zeitschr. f. Morphol and Anthropol.* XXV.

representing a pre-stage of Man, was still definitely simian, and so gave its genus the name of *Australopithecus Africanus*, the South African Ape. There was general agreement among leading scientists that the skull was indeed that of an Ape, and the name *Australopithecus* is still in common use today. As for an explanation of the many surprisingly human features, some authors insisted that the skull was remarkable only for its persistence of *infantile* features; others argued that *Australopithecus* had in a kind of parallel evolution developed certain characters that just happened to be of human resemblance. Both arguments are in no way convincing. The criticism is, in the first case, that the striking persistence of infantile features is, as such, no explanation, but is exactly the peculiarity that has to be explained; and in the second case, that the massive accumulation of distinctively human characters, so indicative of an evolutionary trend towards Man, is another such peculiarity that needs explanation.

It cannot be denied that the *Taung* skull does not resemble what we would readily call a *human* profile. The flat nose and the projecting jaws give it rather a stark concave contour, very similar to that of the Chimpanzee. However, a flat nose and a certain degree of prognathism exists also in certain human races; moreover, the nose bones approach by their greater width and their still incomplete fusion much nearer to the *human* than to the *anthropoid* condition, and its conspicuous jaw-projection does not even reach the moderate degree met with in the young Chimpanzee, let alone the marked prognathism of the young Gorilla. With this picture of *human* approach, the near-*human* milk dentition fits in well. However, there is a small diastema in the upper jaw between the canines and the lateral incisors which clearly is an anthropoid feature; yet, in differentiation from the Apes, no such diastema exists in the lower jaw. In the size and form of the palate and in the width between the canine teeth the *Taung* child again stands nearer to Man. However, the mandible is much heavier than in the human child, and the first permanent molar teeth are correspondingly large, larger even than in the young Chimpanzee. Their considerable size, in conjunction with the diastema in the upper jaw, has therefore been taken as a clear token heralding a Gorilla-like development of teeth and jaws, and hence constituting definite evidence of the anthropoid nature of the skull.

There are, however, no supraorbital ridges which are already prominent on the skulls of Chimpanzee and Gorilla of equivalent age. In this and also in its approach to the formation of a chin the *Taung* child again stands nearer to Man. Still more significant in this respect is the distinct forward position of the foramen magnum, from which Dart logically inferred the erect posture of the *Taung* species. However, other experts insisted that such an anteriorly situated foramen magnum was characteristic of all infantile Apes, and that in any case the *Taung* child would herein lag far behind the modern human child.

Still, from the length of the brain cast it appears that the skull was of a longish shape, and since dolichocephaly is a *human* feature, in contrast to the round-headed Apes, the distribution of weight must have greatly differed from that in the round skulls of the Apes, particularly since in respect of the Apes of a corresponding age the facial skull would already begin to outsize the brain skull. From this point of view, again, the forward position of the foramen magnum puts the *Taung* skull distinctly upon the *human* side.

The brain volume has been estimated at about 600 c.c., hence was much larger than that of the young Chimpanzee and was also somewhat larger than that of the young Gorilla, but it falls easily within the range of the adult male Gorilla, while it reaches only about half the brain volume of the modern human child of corresponding age. This comparatively small volume again led most authors to the conclusion that the *Taung* brain was essentially *anthropoid*.

However, the fact remains that it was of greater length and height than are the brains of the living infantile Apes, and as a further token of progressive development there was a notable steepness of the forehead, and also the arrangement of the junction between the frontal, parietal, and sphenoid skull bones approached to *human* characteristics. Still, it was counter-argued that a well-shaped forehead was a feature common to all infantile Apes, and that the *Taung* brain was, after all, of no such exceptional pattern as to justify its essential discrimination from well-developed *anthropoid* brains.

This brief, and by no means exhaustive, account of the *Taung* skull may suffice to demonstrate its perplexing mixture of anthropoid and human features, which makes it understandable how difficult a task it was for the scientist, when carefully weighing one feature against the other, to arrive at a ready decision.

Proceeding now to the biological method we first consider the criterion of general size and strength, and as the *Taung* skull comes by its absolute measurements near to those of the Gorilla of corresponding age, we may conclude that the species was somewhat taller and stronger in build than was the Chimpanzee. If this were so we should expect the *Taung* child to develop powerful canine teeth after the manner of the Gorilla. Nothing, however, is indicated by the skull that would predict a later development into such a fighting Ape. On the contrary, while in the young Gorilla the milk teeth are on the whole stronger, and particularly the canines already well transcending the level of the other teeth, and the jaws correspondingly expanding in length and width, we find that the *Taung* child not only does not follow suit in all these respects but even lags well behind the dimensions of the Chimpanzee. This is the more significant inasmuch as his comparatively large permanent molars seem to suggest that in his mature life he will develop strong and heavy jaws and a corresponding strength in his general physique. The assumption has been put forward that the *Taung* skull might have belonged to a female Ape. Even on such an assumption the anthropoid nature of the skull cannot convincingly be established since the jaws would still be too small and the brain too large to fit into the picture.

From the fact that the skull was discovered in a limestone cave, among the remains of Baboon-like Monkeys, the conclusion was drawn that the *Taung* creatures lived in caves and upon the ground. As R. Broom put it, "the apes lived on the plains and among rocky krantzes of the present-day Kalahari Desert". The question now arises as to how they moved about on the ground. The answer is given by the forward position of the foramen magnum—they walked in an erect fashion on their two legs. If this were so, that is if they were an erect-walking species of over-Chimpanzee size, yet without any fighting weapons developed within their physical organization, the conclusion is cogent that they stood on the *human* side, irrespective of the fact that no fashioned tools have been found on the *Taung* site. However, the evidence of violent cracks in the skulls of the Baboons indicates that the skulls have been broken up deliberately for extracting the brain, an operation which can only have been done by the blow from some tool.

As long as the *Taung* skull was a solitary find, its immaturity clearly stood in the way of any definite classification, and the biological method, sponsored in this book, and which strongly pointed to its human nature, was far from being accepted by morphologists. So, when in my paper "The Taungs Puzzle" (*Man*, 1934) I challenged the Ape-theory, nobody would listen. Even after the discovery of many more skeletal remains of young and adult specimens with jaws and teeth of strikingly human appearance and also with hips, legs, and arms of clearly human structure and proportions which made the upright gait of the *Taung* species a certainty, the Ape-theory still held its ground. Here again, as in the *Pithecanthropus* case, the ignominiously small brain was taken as definite evidence against human classification —although a smaller brain than that of *Java Man* should have been expected in a species living some considerable time before the other, and to that extent nearer their anthropoid ancestry.*

* While until recently the age of a fossil has been exclusively estimated on the evidence of the geological stratum in which the fossil was situated in conjunction with associated fauna and tool-type, there are nowadays also methods based on geochemical and radioactive tests in use for dating, with results frequently differing considerably from the old estimated *Zinjanthropus*, for instance, the oldest as yet known tool-making hominid whose age is usually dated as lower Pleistocene, that is, perhaps 750,000 years ago, has now on the potassium-argon test been estimated as old as 1,750,000 years. If this estimate is correct, the antiquity

It was only within the last few years, after L. Leakey found in the old Olduvai Gorge in Tanganyika, Africa, direct evidence of chipped pebble stones in association with fossil remains apparently closely related to the *Australopithecinae* that with this discovery of *Zinjanthropus*, the "tool-maker", a change of wind occurred over the Kalahari Desert. Today there is a growing number of scientists who would agree that *Australopithecus* "perhaps passed over the threshold of hominization"; others even group him definitely with the Hominidae, that is, with true Man.

The logical consequence should be to discard the name *Australopithecus* which, after all, is a name denoting an *Ape*, and instead to call him after the site where he was first found: *Taung Man*. As such he would represent the whole genus known now by the collective name of the *Australopithecinae*, comprising both the South African *Taung* species of smaller size and slender build and the *Paranthropus* of larger size and heavier build, as also the heavy *Zinjanthropus* of Kenya, and perhaps also the skeletal remains found at Korotoro in the Sahara.

After *Taung Man* and *Java–Peking Man* the next well-documented stage of human evolution is represented by *Neanderthal Man* who lived in the time of the last *Glaciation*, that is, perhaps 300,000 years after *Peking Man* and about 100,000 years before our own time. He was definitely an advanced human stage with a brain volume ranging up to 1600 c.c. and in the possession of well-designed stone tools. In his anatomical structure, however, he was in many respects of conspicuously simian-like appearance such as is shown by his low forehead, his bulging eyebrows, and the absence of a chin. Since fossils of his type have been found widely distributed over the earth, he must have been

---

of Man would go much further back than has generally been assumed. However, the validity of the test, taken on volcanic material surrounding the bones, has yet to be proved. Leakey's newest discovery in the same Olduvai Gorge of a number of skeletal fragments attributed to an about 4 feet tall, erect-walking, tool-making hominid species, named *Homo habilis*, that is, skilful Man, is even believed to be still older than *Zinjanthropus*. Here again we have to wait for further investigation and evaluation.

representative of a genuine evolutionary stage of humanity, whether or not modern Man developed in direct lineage from him. It was only when the Ice Age was giving way to a warmer climate that, with *Aurignacian Man*, the modern type of Man, *Homo sapiens*, entered the scene.

The long persistence of simian-like features in the skulls of fossil Man has always perplexed the students of palaeontology. Attention has long since been drawn to the observation that "the difference between *Neanderthal Man* and recent Man was much more strongly revealed in the structure of skull and jaws than in the configuration of the limb bones". From the standpoint of this book the "low niveau" of archaic human skulls is easily understood when we consider that it was in the first place the limbs upon which the human principle turned for the sake of perfecting tool-use, and here it was mainly the lower limbs—hips, legs, and feet—that were initially subjected to modification in the direction of upright gait, while the upper limbs were already of an organization well suited for tool-use. On the other hand, the skulls would not likely be affected in their *facial* parts as long as the diet did not greatly change and the mastication of the raw food still required heavy jaws and strong teeth; and in their *cerebral* parts the skulls would only enlarge in height and size when the brain, and here particularly the frontal area, began to grow.

There is also some perplexity among palaeontologists over the question of how long it may have taken Man to develop from his anthropoid ancestry into true Man. Our theory indicates that it must have been a rapid transition, since it was simply the change from the animal to the human principle, as implemented by the systematic use of artificial tools, that created Man. Therefore it was in a comparatively short time that the organs immediately concerned with the new scheme of tool-use came to undergo the transformation from the anthropoid to the human form, as is evidenced by the features of upright gait which were already being impressed upon the *Taung* skeleton.

Another much-ventilated question refers to the sequence and the lineage as yet observable in the evolution of Man from the beginning of the Pleistocene to the present time. It can now be safely said that Man already existed in the first half of the Pleistocene, that is, half a million to one million years ago, in widespread parts of the earth and in various more or less divergent forms—as this is evinced by the *Australopithecinae* in Africa, the *Pithecanthropidae* in East Asia, their contemporaneous *Heidelberg Man* (of whom a very massive yet essentially human lower jaw was discovered at *Mauer* near Heidelberg, Germany), and perhaps also by some skeletal remains found in Ternifine, Algeria.

However, in spite of the many varieties which may have been due to isolation and different environment, Mankind as such was never split up in itself, but was basically of one and the same general type, as which it continued on through the second half of the *Pleistocene* in its various forms. *Neanderthal Man* was apparently the most characteristic and most dominant of those forms. He seems to have rather abruptly entered the scene just before the last *Glacial*, and to have as abruptly disappeared from it at the end of this cold period, when he made room for the modern type of Man. Fossil remains approaching the modern human type have also been found of an earlier geological time, such as the skull fragment discovered at Swanscombe (England), perhaps of the late *Second Interglacial*, and the current view is to the effect that *Homo sapiens* goes further back than *Neanderthal Man*, having developed independently of him. However that be, *Neanderthal Man*, whether a sideline or a central line, was in any case a member of the same human family as modern Man, and for over 50,000 years its dominant type.

GEOLOGICAL TIME SCALE

	Years	Period	Prehistoric Man
Upper Pleistocene	c. 10,000	Postglacial	Neolithic Man
		4. Glaciation	Aurignac Man
	c. 100,000		Neanderthal
		3. Interglacial	Man
Middle Pleistocene		3. Glaciation	
	c. 200,000		
		Great Interglacial	Swanscombe
	c. 400,000		
		2. Glaciation	Peking Man / Java Man
	c. 500,000		
Lower Pleistocene		1. Interglacial	Heidelberg Man
	c. 600,000		
		1. Glaciation	Australopithecinae / Zinjanthropus
	c. 1,000,000	Villafranchian	
		Pliocene	
	c. 15,000,000		
		Miocene	
	c. 30,000,000		
		Oligocene	
	c. 45,000,000		
		Eocene	
	c. 70,000,000		

This table shows a sequence of geological periods. The dating, after various authors, is, at most, only approximate.

Part Four

MAN'S PLACE IN NATURE

CHAPTER 18

Man and the Ape

IF IT is agreed that Man owes his distinctive rank and greatness to his specific principle of evolution, one that dissociates him essentially from the animal class, the idea that he "came from the Ape" will at once lose its painful sting. Man may even take pride in having consistently pursued a course of evolution in which he broke away, and steadily widened the distance, from the animal realm. Yet in his self-assertion he should not go so far as to speak of his former kinship in a derogatory way. Phrases such as picturing the Apes as the "distorted copies", or the "ridiculous caricatures" of Man, only indicate that the Apes' true position in Nature and their true relation to Man are not properly understood.

If we judge the Apes by animal rather than by human standards, our verdict will be different. We then come to realize that the Apes are outstanding among the rest of Animals for their possession of a most elaborate brain and a correspondingly high degree of alertness, sagacity, inventiveness, and adventurousness. This is all to their credit, even if they do not match the grade of human intelligence.

To look upon the Apes as "transformed" or "degenerate" Men would imply that they once actually stood on the *human* side and by some misfortune had fallen back into "brutality". Such an unorthodox conception of the Apes as representing creatures which were formerly *human* beings is not so far remote from the superstitious beliefs of primitive peoples, in which the Apes are actually figured as a freak sort of human beings. Thus the Malayan word Orang-utan literally means "Man of the woods".

Strangely enough, similar ideas of a human strain inherent in the Apes have been put forward even on scientific grounds. If the Common Ancestor, the argument ran, still lacked the overlong arms and hands, the inverted grasping feet, the tusk-like canine teeth, and other disproportions characteristic of Gorilla, Orang, and Chimpanzee, then his general appearance must have been incomparably more *manlike* than it is with the present Manlike Apes. Accordingly, these Apes must have lost their original greater manlikeness due to an adverse evolution. This in itself was quite a logical conclusion, but was interpreted as to mean that the Common Ancestor would by virtue of his greater manlikeness already have stood on the *human* line—a line that Man simply followed up in the course of his evolution while the Apes forfeited all their promising humanity by drifting away from the *central* human road into a side-road of increasing "brutalization".

In other words, it was no longer Man who came from the Ape, but conversely, it was the Ape who came from Man, and old Darwin was wrong.

The fallacy of the argument lies in the tendency to interpret the higher degree of *manlikeness* existing in the Common Ancestor as a criterion for his *humanity*, as if a greater accumulation of manlike features would grant the Animal *human* status. One did not heed the obvious fact that the term "manlike" only refers to a certain *resemblance* to the human form, and does not mean nor imply identity. Thus we call the Apes, because they *resemble* Man in many respects, the "Manlike Apes", and are well aware of the fact that in spite of all their outstanding resemblance to Man they are essentially "Animals", not "Men". Manlikeness, therefore, is essentially an *animal* attribute, and from this interpretation it is evident that even an accumulation of manlike features cannot convert an Animal into Man. Just as the Chimpanzee is for all his great manlikeness not any more "human" than the less manlike Baboon, so is the Common Ancestor for all his supposed still greater manlikeness not any more "human" than the Chimpanzee. It is only through the operation of the human principle that the manlike form is converted into the human form and since the animal principle is opposed to the human principle, it follows that the Animal, however great its manlikeness, is by its own principle of evolution prevented from attaining the "human" status.

If it is rightly understood that the Common Ancestor stood essentially on the *animal* side, it will become clear how great a mistake it was to argue that Man followed in his evolution "the main or central road" running already in the *human* direction and thus remained (what the Common Ancestor was) "human", while the Apes strayed away from the central human line into a side-road running back into animal quarters. From the standpoint taken in this book, it was, conversely, Man who broke away from the central line, the animal line, and thus became Man, while the Apes continued to follow the old line, and thus remained Animals.

As the Apes are not "transformed" or "degenerate" Men, neither are they the victims of "Nature's unsuccessful attempt to make Man". Here all the blame is laid at the door of Nature herself. Yet, if it were ever Nature's ambitious intention to make Man, the very existence of Man would prove that she was indeed most successful in her venture. Nor is the argument valid that the Apes had missed their chance of becoming Man by some kind of ill-luck. Their subsequent course of evolution clearly indicates that they never had such a chance. Apparently it is normal in the universal process of evolution that only some single species are able to "move up" to a higher plane of integration, while all the rest, although bearing their full share in developing the physical requirements for the upward step, are left "behind".

In conclusion, the Ape is neither a sort of degenerate or brutalized Man; nor is Man a sort of transformed Ape, for the simple reason that he is no longer an Ape. Phylogenetically, both hang together by the bonds of blood relationship; biologically, they are separated by the unbridgeable chasm of opposite evolutionary principles.

From this point of view it follows that the Linnaean system of classification in which Man was placed with the Apes in the Order of the Primates requires urgent revision.

G

A new scheme based on genealogical considerations has recently been set out by the American palaeontologist G. G. Simpson, and since it is widely used by scientists, it may briefly be dealt with here. In his scheme the Order of the Primates is divided into two Sub-orders, the *Prosimii* and the *Anthropoidea*, the latter comprising the Monkeys and also the *Hominoidea*. The family of the Hominoidea is again sub-divided into the *Pongidae* or Manlike Apes, extinct and recent, and the *Hominidae* or Men, fossil Man and *Homo sapiens*. As one can see, the term *manlike* is here used twice, first, in the term "anthropoid" (derived from the Greek) and again in the term "hominoid" (derived from the Latin), and since the term "anthropoid" is referring also to the Monkeys, it seems to be used as indicating a lesser degree of "manlikeness" than the term "hominoid" which, referring to both the Manlike Apes and the Hominidae, or Men, seems to be used as indicating a higher and highest degree of manlikeness.

This double version of "manlikeness" may well have served the purpose of finding convenient headings for sub-divisions, yet is most unfortunate, not only because the term "hominoid" or *manlike* is easily confused with "hominid" or *human*, but above all, because the Hominidae, or Men, are in their classification as Hominoidea reduced to *Manlikes*, or Apes.

When Linnaeus coined the name *Primates* as denoting the first and highest Order of the Mammalian Class with Man in it, he had, no doubt, the greatness of Man in mind. But even if Man, as should be done, be removed from this Order, the name Primates may still be in place for the Order, in appreciation of their highest representatives, the intelligent Manlike Apes.

In the following two tables of genealogical classification in which both the derivation and the separation of Man from the animal class is emphasized, I have tried to modify Simpson's scheme according to the theory advanced in this book.

It stands to reason that these two classificatory tables are only meant to serve the purpose of a general and rather simplified scheme outlining the dominant phylogenetic relationships of the Primates between themselves and to Man. In reality, things are

TABLE 1

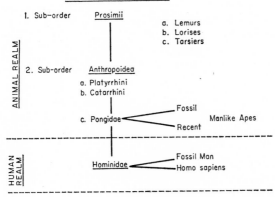

TABLE 2

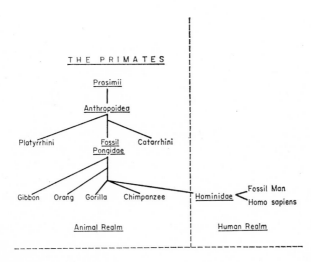

most complicated, and much depends on our ability of determining in the scheme the proper place for the many fossils which seem to have some more or less close affinity with the anthropoid stock and with Man.

As mentioned before, the Gibbons are for their still primitive organization widely believed to form a separate branch of the anthropoid tree, dating possibly back to oligocene times. Another belief is that the miocene and pliocene Apes of the *Dryopithecus* group have given origin to the three great Manlike Apes of which the Orang may have branched off at some earlier time. On the whole, a proper genealogical classification of the Apes and of Man is for the time being still most problematic, and every author seems to have his own views about it.

CHAPTER 19

The Human Level of Integration

FROM time immemorial Man has been led to believe in his "higher destination", not so much for his brilliant achievements in material things, his wonders of technology, but foremost for his profound supremacy of mind, as it is so impressively evidenced in his works of science, art, philosophy, and religion. The old dogma, therefore, not only lifted Man upon a "separate place in Nature", but also claimed for Man a place "outside and above Nature", in contradistinction from the Animal that had its place "in and below Nature".

The Darwinian school, when lining Man up with the Animals, laid the axe to the roots of the old dogma. But whilst they were wrong in denying Man a "separate place in Nature", they were right in refuting the dogma's second claim that Man, again in signal distinction from the Animal, should occupy a place "outside and above Nature".

According to our theory, Man owes his unique existence, and all that evolution has brought him, to his exclusive principle of body-liberation which bestowed upon him the gift of *freedom*. Two spheres of freedom can be discerned : his *physical* freedom freeing him physically from the bondage of his body, and his *spiritual* freedom making him mentally independent of the strings of body-compulsion.

In the sphere of *physical* freedom, his evolutionary principle enables him to adjust himself to the conditions of environment by artificial means and so frees him from the Animal's compulsory scheme of adaptation by bodily means. Fundamentally it is the same ends of adaptation to Nature that are pursued, though in different ways, by both the human and the animal principle, except that Man, by his essentially different and greatly superior artificial devices, attains those ends in an incomparably higher degree which allows him a free choice of environment, natural or Man-made.

It follows that in his physical freedom, in spite of all his remarkable achievements of artificial adaptation, Man stands all "within and below" Nature, and hence would have to pay dearly for any offence against her laws. Thus he may well be able to adapt himself to travelling through the air by artificial means; but if his aeroplane is not adjusted to the specific and varying conditions of the air as adequately as are the body-grown wings of the Bird, he is certain to crash in disaster.

In the sphere of *spiritual* freedom there is again evidence of Man ultimately working towards "natural" ends. On the surface, it is true, it would seem as if in his deliberate suppression of life-preserving instincts and desires, in his scientific, moral, and aesthetic attitude, he would transgress the laws of existence; at bottom, however, we find that it is in pursuance, directly or indirectly, of the same natural ends of adaptation to Nature that the blind instincts are here superseded by conscious idealistic motives.

What with the Animal is sheer "curiosity", namely, an instinctive desire to know whether some strange-looking object has, or has not, any practical relation to its existence, so as to want it or to disregard it, or to keep clear of it, in the case of Man this passion for inquiring what is existing and going on in the world is in the fields of pure science sublimated into the pursuit of truth and knowledge for their own sake. Such academic research must, with the growing insight into the general and special conditions and relations on earth, have an increasing influence upon the practical life-sphere of Man and so become in the end subservient to immediate needs and interests. Indeed, without the thousand years' unceasing search for the theoretical foundations and principles of mathematics, physics, and chemistry, Technology would never have reached the tremendous height of its present-day achievements in the adaptation to, and exploitation of, any kind of environment. Even abstract philosophy, although primarily concerned with fundamentals and the transcendental significance of things and phenomena, eventually meets the practical sphere of Man when it undertakes to supply him with supreme regulative precepts for an idealistic conduct of his affairs.

So has Morality its practical aspects; and here again we find Man ultimately working towards natural ends. In the case of the Animals, the maintenance of the species is secured by deep-rooted and blind instincts; with Man, the same function rests on his free and conscious volition, that is, on his moral sense. Although it would seem as if in his moral freedom when his instincts are overlaid by idealistic motives, Man would pass *beyond* the sphere of adaptation, the point is that even in the conscious sacrifices and moral efforts there prevails a significant relation to life and its natural demands. Any altruistic behaviour such as parenthood, love, care, charity, indeed any humanitarian enterprise to establish an ideal state of brotherhood among Men—although all such effort may be prompted by the abstract idea of good—must in effect promote the maintenance and the welfare of the human species.

Aesthetics, too, cannot be denied its practical significance in human life, quite apart from the spell of enjoyment and enrichment that it brings. Whether or not Darwin was right in assuming that there existed sexual selection among Animals in response to a certain sense of beauty, Man, in any case, has become able to form of himself a distinct beauty-ideal on aesthetic lines. With him, then, the instinctive function of sexual selection would be apt to be overruled by his free and conscious decisions working towards the same natural ends. His idealistic picture of human beauty, so impressively mirrored in the statues of the old Greek sculptors, may even inspire him to bring his physical condition as near as possible to that ideal state of all-round fitness by appropriate measures of eugenic cultivation such as sport and gymnastics. We know that with Animals it is the constant pressure exercised upon them by their struggle for existence that keeps

their physique at the height of fitness. Thus, when Man systematically trains his body by different kinds of athletics and physical
culture, he has in mind the same ends of physical fitness. What
then in the higher interest of the species is, within the animal
realm, enforced through the operation of blind instincts and desires is, in the human sphere of spiritual freedom, lifted upon a
higher plane, and is achieved here through the operation of
conscious ideals. This may not be the whole story of Man's
idealistic aims, and other points have to be considered as well,
such as those that will be dealt with in the last chapter of the
book. However, from what we have seen so far of Man working
ultimately, on a higher plane, towards the same natural ends of
adaptation as the Animal, it follows that in his spiritual freedom he
also stands "in and below Nature", and has to obey her laws if he
will not come to harm. For instance, a moral code when based on
erroneous conceptions biological, philosophical, or political, and
thus falling short of the Animal's species-preserving instincts, must
in the end become deleterious to the existence and the welfare of
the human species.

This, then, is the position of Man in relation to the Animal: he
stands outside and above the Animal, but inside and below
Nature. If we consider the implication that his distinctive evolutionary principle is for all its novelty and heterogeneity but a
"natural" principle grown in the same fertile soil of Nature as the
animal principle, we should not even expect Man to be able to
transgress the boundaries set by Nature, but must suppose him to
be linked, as part of Nature, biologically and ecologically with the
general and special living conditions on earth. However, even if
we have to take Man down from the high pedestal erected for him
in former times for his alleged independence of Nature's laws,
this does not in the least mean to belittle his actual possession of
freedom. While the Animal is in its ways of life strictly directed
by, and confined to, the limited means of its bodily structures,
Man in his physical and spiritual freedom enjoys an unpre-
cedented and unparalleled power of shaping his environment and
his conduct of life by his own free volition and by making, for his

maintenance, the greatest possible use of all resources and materials offered to him by Nature.

Still, great as Man be in the spheres of freedom, it should be clear that his possession of freedom, physical and spiritual, stems from the virtues of his principle of body-liberation, and hence can only have reference to his liberation from the limitations and dictates of the body. Even if, as shall be later discussed, there are certain facts which give the impression that in his spiritual freedom Man may actually reach *beyond* the circumscribed zone of mere adaptation, he will not reach beyond the boundaries set up by Nature, as the old dogma would have it. Yet in his spiritual sphere, in which the blind and coercive instincts and desires are superseded by free and conscious idealistic thought, volition, and feeling, the human conduct of life, already highly superior to the Animal in its sphere of physical freedom, is further lifted up on to a still higher level of existence, and thereby proves most impressively that Man, although standing in and below Nature, represents on earth a new category of his own, one in which Life, now based on freedom, reaches a *higher plane of integration.*

CHAPTER 20

The Natural Destination of Man

LONG before science came to investigate the natural conditions and circumstances connected with the origin and the evolution of Man—at a time, that is, when Man only intuitively and vaguely grasped the idea of his "higher" destination—his conduct of life had already been greatly inspired and influenced by those great teachers of religious and philosophical systems who laid down moral laws and ethical maxims for guiding him towards a higher state of perfection in which the lower instincts and emotions were superseded by idealistic reflection and free volition. It was not, however, until the closing half of the eighteenth century that one began to look upon human life and its deeper significance from *natural* points of view. Kant, the great German philosopher, who already proposed a theory interpreting our Solar System in the light of natural or mechanistic evolution, was to my knowledge also one of the first to think of the origin of Man in terms of a natural transition from a *quasi*-animal condition into the human stage. In his own words, "the exodus of Man from the Paradise . . . was nothing else than the passage from the crudity of a merely animal creature to the stage of humanity, from the leading strings of the Instinct to the guidance by Reason, in short, from the guardianship of Nature to the state of Freedom". He thus regarded it to be the natural task of Man to take his destiny in his own hand, and under the guidance of his reason and in the consciousness of his freedom to lay the foundation of a human society of the highest possible perfection. To quote him again, "the greatest problem the solution of which Nature has imposed upon Man is the setting up of a law-administering civil society".

Therefore, when Man is striving to resolve this greatest problem, he acts in accordance and in co-operation with Nature. That is to say, the idealistic aims of Man towards the perfection of his species are, as the philosophical expression of a deeply felt biological need, in conformity with the natural ends of his evolution.

With the advent of the theory of evolution when the intrinsic genetic connections between the various living things, with the inclusion of Man, became apparent, the same basic conception of the *natural* existence and development of the human species gained fresh momentum, and was eagerly taken up by a new school of thought, *Sociology*, founded by the French philosopher A. Comte, and further developed, on Darwinian lines, by the English philosopher Herbert Spencer. Their point was that, judging from the phylogenetic unity of all living beings on earth, Man, as an integral part of the living world, must have been submitted to the same principles as the Animal in the course of his evolution. Hence his present "cultural" stage of civilization, with all its tremendous technical and mental achievements, and its advanced social and political institutions, was no longer taken as being opposed to Nature, as it was formerly believed, but was now held to be the straightforward outcome of "natural" evolution. Accordingly, the "History of Civilization" was claimed to be identical with the "Natural History of Mankind".

The start of Sociology was brilliant enough in its methodological integration of the unique process of civilization into the universal process of evolution, as well as in its implication that Man had now become conscious of his own evolution as part of this universal process. This spectacular event of "evolution transcending in Man the threshold of consciousness" was rightly emphasized by sociologists as the great turning-point in human history, when "Man awoke from his thousand years' sleep of Animal-like instinctive existence", and from now on was able to supplant the former blind automatism of his evolution by conscious and rational direction.

However, to take his evolution in his own hands and control, and to co-operate in this task intelligently with Nature, it is not enough for him to know the mere fact of his natural evolution, and to recognize the great significance that this knowledge would have for his future development. What he most urgently needs to know is the principle itself that lies at the root of his distinctive evolution—the principle that he had so far followed unconsciously and now was called upon to follow consciously.

Sociologists, it is true, were most successful in fitting the sublime humanitarian ideas and promises of our great philosophers harmoniously to the conception of the organic unity of the living world. But they failed to realize that the uniqueness of the human process of evolution, as a natural event, must have been due to an equally unique principle that was at the root of the process. Instead, they contented themselves with the superficial argument that the same factors that were found to be operating in animal evolution must naturally hold good also of human evolution. In this conception they were the more prejudiced, inasmuch as the comparison of the appearance of modern Man with Ape-like *Neanderthal Man* seemed to prove that Man was in his "cultural" evolution well on his way towards a higher *organic* type—similar to those instances observable to animal evolution where lowlier forms developed into forms of higher organization.

Thus human evolution was equated with animal evolution, and for its final stage a "Superman" was stipulated who was conceived as being physically and culturally as remote from present Man as present Man was remote from the Apes and the Monkeys.

Such being the state of affairs, with the Darwinian Principle of the "survival of the fittest" figuring as the natural agent of human evolution and with a "Superman" set up as its natural goal, it is hardly surprising that a number of writers and politicians would for the sake of the creation of Superman even recommend a speeding-up of the selectionist process by the deliberate extermination of what they think to be the physically or mentally unfit. F. Nietzsche, the celebrated inventor of the zoological "Superman", set the fashion here. Meanwhile, P. Kropotkin had warned us of the fact that within the animal world there was not only preying and competition but mutual aid and co-operation as well. Also Spencer's suggestion that in human evolution the centre of gravity had shifted from the "organic" to a "superorganic" plane was not ignored. Still, with the actual principle of civilization not yet being properly grasped, the whole situation was bound to remain ambiguous and open to unscrupulous consequences, and it was left to Hitler-Germany to carry Nietzsche's Superman-ideology into ruthless practice by mass-sterilization, mass-poisoning, and mass-murder in gas chambers.

Within recent years, in reaction to the stirring experience of the Second World War, the necessity of a rational control of human evolution has gained new ground. It was now recognized that the recent advances in the knowledge of both the structural elements within the *material* world and the psycho-social and genetic elements within the *living* world must lead to a new pattern of human thought and behaviour and bring on a new phase of human evolution in which natural selection was subordinated to the operation of psycho-social ideas and values. "Man is embarked on the psycho-social stage of evolution", we read in Sir Julian Huxley's book *The Humanist Frame*, 1961, involving "radical" change in the dominant idea-systems", and an organization which is "ideological instead of physiological or biological".

Here, at last, an attempt is made by a leading Darwinian scientist to bestow upon Man an evolutionary principle essentially different from that governing animal evolution. "Man's evolution", he says, "is not biological but psycho-social: it operates by the mechanism of cultural tradition, which involves the cumulative self-reproduction and self-variation of mental activities and their products."

In this thesis emphasis is laid on the contrast between the "ideological" evolution of Man and the "physiological or biological" evolution of the Animal. If the terms "physiological or biological", since they apparently refer to the *bodily* evolution of the Animal, are here logically replaced by the term "bodily", not only would the *bodily* principle of animal evolution be clearly distinguished from the extra-bodily principle of human evolution, but the term "biological" would still be available for both animal and human evolution which are, each of them, obviously *biological* processes. At the same time we would be guarded against the temptation to mistake the new "psycho-social phase" for the true start of human evolution and so drop all the preceding stages of humanity into oblivion. Also we would be guarded against the dangerous tendency to applying the doctrine of the "survival of the fittest" (referring in the first place to *bodily* fitness) unreservedly to Man, as if he, like the Animal, would develop into a creature of higher physical organization.

If we go by our own theory the process of human evolution is, in all its successive phases, a unitary event rooted exclusively in the principle of body-liberation. Hence it is primarily focused upon the development of artificial tools, material and mental, while the body takes second place in its adaptation to tool-use. Therefore, automobiles, aeroplanes, electronic computers, washing machines and antibiotics are the things in line with, and promoted by, the human principle, and it is not only among the technical devices, but equally among the spiritual issues such as sciences, ideologies, political and social constitutions, economics and ethics where the principle of the survival of the fittest enters.

Since all cultural qualities and achievements of Man are derived from the same source of the principle of body-liberation, it thus appears that this principle is itself the mysterious principle of civilization, and that it is merely the cultural level of civilization that changes with the progress of the human principle, not the principle itself. Civilization, indeed, began with the first systematic use of tools and with the first dawn of freedom from body-compulsion and hence is as old as Man himself. This causal relationship between tool-use and civilization is well recognized by some modern writers. So we find in Lewis Mumford's *Technics and Civilization*, 1934, the following sentence : "During a great span of primitive life the slow perfection of stone tools was one of the principal marks of its advancing civilization and its control over the environment."

If the principle of body liberation is accepted as the unique principle underlying the unique process of civilization, the great cultural movement of Mankind in its intricate ways will stand out more clearly against the blurred background of human history. Under its guidance we may not only be able to follow the cultural movement backward to its early stages; but from the mode and direction of its previous stages we may even draw some tentative inferences upon its future course.

Apparently it was not a smooth path that Mankind was destined to travel in its evolution. From what History tells us of unceasing struggle and warfare, sufferings and unrest, we may conclude that there has always been "blood, toil, tears, and sweat" since the time that Primeval Man, himself struggling for his existence, started upon his adventurous career. If we take too narrow a view of the march of human evolution by concentrating upon its "historical" period only, we must be prepared to get a cultural picture of Man that is anything but propitious. No wonder that not a few philosophers, disillusioned by those ignominious records of human imperfection, have given voice to a most pessimistic outlook on the prospects of the human species. Technical progress they would not deny; but they insist that not the slightest progress of any significance has been made in Man's spiritual sphere. Some even would believe that civilization as such was to be blamed for the cultural stagnation and the spiritual deterioration of Mankind, and Rousseau's warning of "Back to Nature" is still in the air.

In the wider perspective of the human movement, however, things look different. In the dimness of the remote past the cultural process of civilization sets out upon its enterprising beginning and the first evidence of small but definite progress that we find is the crude stone implement of early Man. *Peking Man* was already fairly skilled in the making of stone tools. He was a hunter of Animals and knew the use of fire. It is difficult to draw any direct conclusions on his social life from the scarce evidence of early Man. However, if there is evidence of the use of fire and of numerous remains of hunted Animals, we may take it as a fair guess that Man at that time already lived in ordered social groups or communities.

A valuable informatory clue to the social life of early Man would be provided if we could find evidence of ceremonial acts, such as the burying of the dead. So far, however, no such evidence is available until the time of *Neanderthal Man,* who already showed greatly advanced brain development, and must be supposed to have had his intellectual and social capacities equally enlarged. He certainly had a very hard time when he sought shelter in caves against the severe rigours of the Ice Age, and it may have been the inclemency of the climate that largely retarded his further development. So when *Aurignac Man* migrated into his territory, this newcomer had already elsewhere reached the anatomical type of *Homo sapiens,* probably under more favourable climatic conditions, and was also well on his way towards modern Man in his social and spiritual life. Palaeolithic craftsmanship came with him to its peak, and it was he who, in the so-called *Magdalenian Period,* decorated his cave walls with those beautiful polychrome pictures of realistic artistry and which still strike us with wonder.

From then, in the *Neolithic Age,* the process of civilization advanced steeply. Revolutionary novelties, such as agriculture and the domestication of Animals and the arts of weaving and of pottery, sprang up, and for the first time metal was also used as material for tools and ornaments. So great a technological and cultural progress must have been concomitant with mental progress, and in the old Babylonian civilization, starting perhaps 8000 years ago, we meet already with large cities, administrative legislation, and a well-developed system of writing. Over the old civilizations, particularly the Greek and Roman civilizations, the road of the cultural development of Man steadily approached the modern level, when with Galileo the conception of modern science was born, and in its train the stage was prepared and now has been reached when scientific Man has become aware of the phylogenetic unity of all life on earth and the naturalness of his own evolution. Thus, it was not just by chance that unconsciousness of evolution turned into consciousness. With advancing science the day was bound to dawn when the process of evolution came within the scope of science. Therefore, the critical event when Man became aware of his own nature and evolution was in the very line of the evolutionary principle which conferred material and spiritual freedom upon him and now urged him to be guided no longer, as was the Animal, by the dictates of blind instincts and desires, but henceforth to follow the precepts of his own Reason, and by taking his further cultural development under his own control, to act on his own responsibility in harmony with Nature.

In other words, the planning and shaping of his evolutionary future, the mastery of his destinies, is in the order of the *natural destination of Man.*

In this sphere of conscious idealism and volition in which the compulsion of blind instincts gives way to the free consciousness of sublime motives, the human principle reveals itself in a new light, in that it strikes here, in its forms of spiritual freedom, a distinctively "ethical" note. It is indeed in this very light of its *ethical* implications that the greatness and significance of the process of human evolution finds its supreme expression.

The basic principle of body-liberation remains the same in all successive stages; but gradually spreading from one department of human activities to the other it subsequently succeeds in embarking Man upon a style of life in which the maintenance and welfare of his species is no longer supported and directed by a fixed set of instincts, but is entrusted, in increasing degree, to the free guidance of conscious ethical motives. Man has thus to face a new situation in which Nature no longer holds her protecting hand over him, but now charges him with the heavy burden of his own *responsibility* and obligations to himself, to the human community, and to Nature. His deprivation of instincts, Nature's means of control, is in itself a great but precarious achievement, and often enough has led to fatal errors. However, since there is good reason to believe that the conscious control of his destinies is in the very line of his evolution, we may feel confident that some day Man will "come of age", and then attain to such degree of "maturity" as is required for a truly ethical conduct of life in harmony with Nature. In this final stage of his evolution, in which life is deliberately based on ethical consciousness, Man will be "extra-bodily" adapted to Nature as perfectly as the Animal is "bodily" adapted. Then the great movement which converted Man from an irrational into a rational being, from an instinct-ruled life into a free ethical existence, from a slave into a master of his destinies, will have reached, *on an ethical basis*, the ideal equilibrium of adaptive integration that we find with the Animal on an instinctive basis. Then, in terms of biological evolution, the stage of "Superman" is reached.

*

Alas, we seem to be still utterly remote from such an ideal state of equilibrium. The present situation of world affairs even gives the impression that the world is right "out of joint", and that no prospects of an early solution to the many unsettled problems and conflicts within Mankind are in sight. Still, taking a long view on the process of human evolution we note that it is no more than a paltry ten thousand years that separate us from the Stone Age when *Neolithic Man* was struggling hard in developing the material and spiritual foundations of our present era. In this perspective we come to do fuller justice to the great cultural progress that Man has already made in so short a geological time, thanks mainly to his enormous achievements in the technical fields. What precious gift modern technology has given with its time-and-space-overcoming devices of traffic and of communication is best realized when we consider that those devices have broken through the old barriers of racial and national prejudice and segregation. The wide diffusion of comradeship and thought is well exemplified by the international expansion of traffic and trade, exhibitions and sport, broadcasting and film, science and art, and many special institutions serving mutual aid and solidarity in the economic and social fields—all of them distinct tokens of the awakening of a world conscience.

Yet the high cosmopolitan ideas of the great humanists of the nineteenth century transcending frontiers and ideologies are today less respected than ever and frequently are even looked upon as antiquated and impracticable romanticisms. It is true that new waves of national, racial, economic, and ideological self-assertion are spreading over the globe which are apt to augment friction and tension between classes, races, and nations, and subsequently to lead to fresh clashes and conflicts which again must be the more ruthless the more effectively the fighting weapons gain in destructive power. All this may seem to indicate that modern technical progress would virtually counteract an ideal state of Mankind rather than promote it. Nothing, however, would be more unjust than to blame technology for being exploited by Man for destructive or subversive ends—the aeroplane for carrying bombs, the wireless transmitter for diffusing deceitful propaganda, and the knowledge of nuclear fission for producing the H-bomb. After all, technology can only offer Man its highly developed contrivances, and it is for him to choose which use to make of them.

Technical progress has now made it possible that the wars, formerly a matter of local strife and hence of minor significance, are now on the global scale and are bound to shake the whole of Mankind, and by means of their ever-growing devices of destruction to end in mutual devastation and mutilation, and to lead to such appalling repercussions, physical, spiritual, and economic, that the whole structure of human life must be disrupted. It is impossible to imagine what consequences a war, with mega-H-bombs in action, might have on the fate of Mankind, when possibly hundreds of millions of lives—human, animal, and plant—would be wiped out. Still, even such vision of an unimaginable magnitude of destruction, with no victor on either side, may have the salutary effect of encouraging Man to try with every possible means to prevent such suicidal war. Ultimately the fear of nuclear war, or in the last resort the deep wounds inflicted by the nuclear bombs, will drive the nations, with or without their will, to unite in establishing a world order in which the major conflicts are settled by reason rather than by force.

Modern technology has also contributed to the welfare of Mankind by providing new means of conquering hunger and disease.

Agricultural machines, fertilizers, and insecticides are doubling and trebling crop yields. Irrigation systems are reversing soil erosion and turning swampy or barren land into fertile soil; antibiotics are helping to stamp out the deadly pests of infectious diseases. But here again, technological progress is not without serious disturbances in its train. Not only are synthetic insecticides apt to show harmful effects on useful insects and other Animals, but it must be open to doubt whether chemically raised food is as healthy as that grown naturally. Furthermore, there is the problem of overpopulation resulting from the control of famine and disease. This applies particularly to those countries in which the population ratio, formerly kept by Nature in balance by way of starvation and high mortality, is now rising prodigiously through the humanitarian effort of supplying these famine-and-disease-stricken countries with food and medicine.

Still, the attempts at improving the living conditions of Mankind as a whole are of rather recent date and any such unfavourable side-effects will in the end be overcome by new scientific research and improved technological devices.

For his great efforts in making the globe a beneficial and comfortable habitat of his, Man has been reproached for ruthlessly exploiting the earth's valuable resources—coal, oil, minerals, metals—which in time are bound to become exhausted. Here again technology will lend a hand, when natural material will be replaced by artificial products, synthetics and plastics, and heat and electricity will be generated by nuclear power.

Last but not least, besides changing the aspects of human relationship between nations and races, modern technology has equally changed the relation of *individual* Man to his community. Machine production is taking the place of former manual craft on a large scale, automation is replacing Man with regard to the operation of machines; electronic computers are installed to do all kinds of clerical work and calculation, mathematical, statistical, and economic. Man will thus be increasingly released of the former drudgery connected with any such work as is now being taken over by technological devices. The drawback here lies in the problems of redundancy and unemployment caused by these labour-saving techniques.

The "Declaration of Human Right" in 1948 is another token of evolutionary progress in the fulfilment of individual Man. Through the modern educational systems, the establishment of a Welfare State, the notion of "material comfort and security for all", and a new kind of relationship developing between employer and employee, the former class distinctions which once produced Marxist ideology are bound to fade away and to make room for an ideology based on humanist principles.

Human evolution is still in full flux, and its path is inevitably beset with new problems and difficulties. So long, then, as the present transitional period is in progress, we must expect the pages of human history to continue to be filled with the blood-stained stories of wars and conflicts, of political, social and economic upheavals, and of ideological clashes and crises—as the palpable sign of Man groping for the right way, his rushing forward and falling back, his ideological errors and ethical failures.

It is in this perspective that we must look upon the turmoil of present-day events, when the globe is on fire and the atomic bomb looms on the dark horizon. Things are apparently in a state of ferment and the whole of civilization is on the move, not only in the immense strides of technological progress but equally in the fields of religion, economics, education, politics; even in the secluded spheres of arts—music, painting, sculpture, architecture. Everywhere we note supreme efforts being made, exploratory, adventurous, towards an unknown destination. Will Man ever reach his goal? As A. N. Whitehead reminded us, Man is "the servant and the minister of Nature. It still remains to be seen whether the same actor can play both parts."

We do not know whether Man will ever rise to the occasion, nor do we even know what Nature, on her part, has still in store for Man. The great earthquake of San Francisco in 1906, which destroyed almost the whole town, still hovers in our memory. If geologists are right in their assumption that we are now living in an "interglacial period", that is, a warm period between two Ice

Ages, we must reckon, in the more or less remote future, with a complete change of environment, when the northern and southern regions of the earth will be covered with thick ice sheets.

Still, we may look with confidence to the future, and trust that science will in the end solve all the intricate problems which are now lying so heavily upon Mankind, including climate control.

Whether or not we believe in the perfectibility of human nature, sooner or later Man will reach his destined goal. Once Primeval Man had put himself under the principle of body-liberation, Man is inexorably and inextricably in the grip of this principle whose dynamic power, so impressively shown in the achievements of technology, will push him further and further up on his way, until Reason has conferred upon him the wisdom to put his own house in order, and to take his destinies successfully in his hands. Technology, with which the principle started, is still the pace-maker, and this gives our "Atomic Age" its profound significance.

CHAPTER 21

Evolution and Metaphysics

ORTHODOX philosophers are looking askance at any attempt to describe human nature, human efforts, and human destinies in terms of mere biology. In their eyes, the only ethical consequence of "Biologism" would be a kind of dull quietism, that is to say, a hopeless state of spiritual stagnation in which "the high tide of life will ebb away"—a state in which Man, left with no ethical aims and no mental wings to lift him "beyond himself" into the spheres of the Infinite and Absolute, would merely "vegetate" in a life of monotony and self-complacency such as would never allow him to rise any more to true spiritual greatness and significance. Yet for all their well-meant exhortations against the misfortune of losing all richness from life, it seems difficult to believe that Man, after having reached an ideal balance of humanitarian co-operation and social integration, should at the same time be deprived of his idealistic emotions and higher aspirations. With every fresh progress of science and thought new problems are bound to arise and to open new vistas for the effort and the enterprise of Man with new and remote goals for which to strive.

Although we may be reluctant to listen to philosophers when they predict a gloomy futurity of spiritual stagnation, we should not ignore their warning voices when they point to the *limits* drawn to any interpretation and evaluation of natural phenomena on biological or mechanistic lines. Biology, they would admit, is extremely useful and indeed indispensable for the scientific investigation into the process of animal and human evolution. But they refuse to believe that biological methodology is competent and

185

complete in itself; that is to say, that it would actually grasp and comprehend the process of evolution in all its aspects so as to give an exhaustive account of the full significance of human evolution; and they demand that under no circumstances should we confuse indiscriminately the biological concepts of science with those ethical values that are inherent in human evolution and lacking in biological concepts.

However, many biologists, and not only "incorrigible materialists", would insist that the *conscious conduct of life*, may it be ever so significant for Man, does not differ in essence from the instinctive pattern of animal behaviour. Both, they urge, serve the same ends of adaptation, only that the conscious modality would lift human life to a higher plane of integration. From biological points of view this may be quite a plausible argument; yet it leaves open some few fundamental questions.

First, the conscious conduct of life is linked with the implication of responsibility. Man, no longer controlled like the Animal by blind instincts, has to manage for himself, by virtue of his Reason, to conduct his life in conformance with his conscience and with his obligations to himself and his species, and by such behaviour raises the level of his existence to a sphere which signifies itself as "ethical" and is as such non-existent in the animal realm.

Secondly, there is good reason to believe that Man's conscious conduct of life is not entirely consumed in the mere function of adaptation to Nature, but in some departments of his physical and spiritual freedom may go *beyond* that function. Suppose evolution has reached in the case of Man, contrary to the Animal, a stage "beyond that necessary to provide for merely bodily existence". This would be an additional asset which, although a natural outcome of his evolution, could not readily be accounted for on purely biological lines.

For example, it appears that Man's cultural evolution is liable to break up all natural development on earth. What is of use or advantage to him he is anxious to cultivate; what is harmful he seeks to destroy. Before long he will have turned the whole globe into his private grounds of household economy—his laboratory, vegetable-garden, power-station, and a park for wild and domesticated animals. Such a deliberate control and utilization of the whole of creation, animate and inanimate, is difficult to refer to mere adaptation to Nature. Rather would it seem as if Man, conversely, is adapting Nature to himself by harnessing whatever he can grasp to his own needs and ends.

Another kind of enrichment of life we meet in the spiritual zone. Man's devotion to science, astronomy, philosophy; his edification in religion, ethics, and mysticism; his passion for beauty in Nature and art—all these supreme qualities of Man are solemnly and harmoniously woven into the texture of his life, and yet are not necessarily required for mere existential adaptation to Nature.

To state, then, that *consciousness* in the conduct of life suffices to account also for the "higher" faculties of Man would mean to concentrate too rigidly upon the evolutionary angle of the natural growth of his accomplishments to the neglect of the question as to whether in certain supreme cultural achievements he may not actually transgress the narrow margin of mere usefulness and adaptation.

Attempts have been made to get round this tricky problem by taking the "higher" achievements of Man in the light of mere by-products of evolution, as it were, "biological luxury". It is difficult, however, to believe that the very highest qualities of Man should be but accidental growth. Nearer to the point, it would appear, are those biologists who speak here of "surplus" achievements of civilization. Since the term "surplus" refers to something that reaches *beyond* the limits of ordinary requirements, they would in effect admit that evolution has here gone "beyond that point necessary to provide for the mere bodily existence of Man".

If it is said that the stage of consciousness lifts the human conduct of life upon a "higher plane of evolution", the question arises what actually is meant and implied by that phrase, and here, indeed, we are putting our finger on the most critical point of biological argument.

That there exist in evolution different planes of organization from lower to higher states is a safe conclusion. To all appearance, the Ape stands on a higher plane of organizational integration in the animal hierarchy than, say, the Amoeba, and it is in keeping with this evidential fact that biologists speak of the "lower" and "higher" forms of life, the latter being distinguished from the former by their "higher differentiation and integration of structure" and a correspondingly higher degree of mental abilities. It has, however, been urged as an objection that the terms "low" and "high", if used as indicating *evolutionary progress*, would contain a value-judgement which does not readily fit into the observation that the lower organizational forms are as adequately adapted to their surroundings as are the forms of higher organization, and that many higher forms have become extinct, whilst lower forms have survived; furthermore, if there were any "progress" in evolution, this was the exception rather than the rule, since the lower types were not only still in existence but were even in the great majority. Some biologists, therefore, would altogether deny true progress in evolution, or at most grant it to human evolution alone.

J. Huxley, in his book *Evolution as a Process*, 1951, when searching for an objective criterion of true progress in evolution, made the point that all progressive features of a *universal* character, such as adaptation, specialization, etc., would indicate only *one-sided* biological advance which as such was characteristic of life itself; whereas true progress was revealed in those features which showed *all-round* advances and thereby raised the upper level of organization and biological efficiency—advances rightly described in biological textbooks as the "great steps" in evolution, such as the step from the unicellular to the multicellular organism, from the Reptile to the Mammal, from the Ape to Man. It cannot indeed be in doubt that those great steps are of a *progressive* kind, if this term is taken in the sense of "higher differentiation and integration of structure", and not in the sense of the higher organism standing in rank *over* the lower one.

There is general agreement among competent biologists that the evolutionary process, far from being caused by some supernatural agency, was a purely *mechanical* affair due, on the one hand, to mutational changes in the hereditary constitution, and on the other hand to the sifting and perpetuating operation of natural selection; and since mutations have been found to occur abruptly and at random, the process of evolution was taken to be essentially a *matter of mere chance*.

The chance-theory of evolution, already suggested by Darwin himself, is with its *anti-deterministic* principle, which excludes any form of design or purposiveness, falling well into line with the *mechanistic* conception of evolution and is, incidentally, also in line with the modern principle of indeterminacy in the physical field. It is, however, difficult to believe that the miraculous ascent of evolution up to Man as its crown should have been the contingent play of an incalculably long series of single random variations, all extremely small, yet all extremely fitting; nor is this difficulty lessened by the evidence of an *immensity* of time that it took evolution to progress from the lower to the higher forms of life.

The difficulty is even greater now that recent research into the basic elements of hereditary transmission has shown irrefutably that changes in the genotype of organisms are only and exclusively brought about by changes in the gene material of the chromosomes, that is, by "mutation". However, scientists are still in the dark about the mechanism by which mutations occur. Experiments in artificially induced hereditary changes, by radioactive and other stimulants, have so far produced mutations only of an abnormal or harmful kind, apparently through damage done to the genes.

One thing, in any case, has definitely been established, namely, that "acquired characteristics" are not inherited as against the opposite doctrine that has been, rightly or wrongly, laid at the door of Lamarck. What really was meant by an "acquired characteristic" has been the source of much confusion. When at the end of the last century the German scientist A. Weismann hacked off the tails of a hundred generations of rats and then found that the young rats still showed their proud tails, he claimed to have proved conclusively the non-inheritance of "acquired characters". But surely, Lamarck never thought of mutilation as being inheritable; he was rather concerned with the use and disuse of organs as an evolutionary factor, in so far as biological need resulting from changed environment would have influenced animal behaviour. Thus in the much-ridiculed instance of the Giraffe sticking out its neck to reach the leaves of the trees and subsequently developing its long neck, it should no longer be supposed that it was simply through stretching it that the neck became longer, but that the Giraffe stretched its neck because this was in its evolutionary trend —just as swift running on its toes was in the trend of the Horse's evolution from a five-toed into a one-toed Animal.

Only very recently, the intrinsic relation between evolutionary trend and environment has been emphasized by the eminent geneticist C. H. Waddington* when he challenged the current theory that random mutation plus natural selection would be sufficient to account for the upward movement of evolution. Animals, he urged, were well capable of selecting *for themselves* their particular environment and thereby would gain a decisive influence on the type of natural selective pressure imposed upon them by the new environment. In his theory *animal behaviour* in its relation to environment is thus taken to have effectively contributed to the course taken by evolution, which in a way is plain Lamarckism, yet in the modern sense that the animal behaviour is conditioned by hereditary factors enwrapped in the chromosomes.

The basic question here is how far is animal behaviour, in response to environment, influenced by hereditary constitution, and how far may it, on its part, itself influence the direction of mutational changes? The question is not readily accessible to experimental investigation; but the possibility of *directed muta-tion* is very much in the mind of modern geneticists, and some promising experiments have already been carried out in this line. The paramount role that the environment has played in the occurrence of mutational changes can safely be inferred from the experiments that Nature herself has made on the large scale. In

* C. H. Waddington, *The Nature of Life*, 1961.

the case of the Whales, for instance, we do not know why those originally terrestrial Animals have taken to a habitat in the seas, but the close causal relation between their taking to aquatic life and changing their hereditary constitution in adaptation to the new environment is clearly evidenced by their transformation into fishlike creatures. That this inner relation has so far not yielded to experimental verification in the laboratory should not be taken as sufficient ground for denying it.

The "great steps" in evolution are characterized by their involving only the few single species which had the fortuitous chance of a changing environment. Thus, the whole class of Fish, although well disposed of spreading, on a mutational basis, into innumerable variations, was by their over-all aquatic specialization practically barred from the next higher "Amphibian" stage with the only exception of the one species of the Lungfish group which, living in shallow and seasonally drying-up waters, had gradually adjusted itself to temporary life on the land by breathing air through lung-like organs and by crawling and clambering about on its stocky fleshy fins, which by their central bony structure and dermal rays are forestalling the limbs of the primitive Amphibians. So when need arose, such a single species was already prepared and ready to embark upon a new principle of organization, thereby reaching a higher level of biological integration and forming a new dominant type in which the new pattern was further worked out into an abundant number of varying forms, one of which was eventually again to start upon a new principle of organization.*

* The Lungfish are generally believed to have given birth to the Amphibians. They are of particular interest as we still see them using their fins—that is, their *swimming* organs—for moving about on the land in a clumsy and strained bid for survival. As an evolutionary result, the fins became fittingly reinforced and developed finally (with the Amphibians) into true *walking* organs. This sequence of genuine swimming organs being transformed into proper walking organs leads us to assume that *walking*, initiated and sustained by environmental stresses, has as such played a significant role in the process, whatsoever was the hereditary mechanism effectuating the transformation in conjunction with natural selection. From a logical point, this sequence could as well have occurred in the reverse direction, the organ being prior to the function, such as would be

So was Primeval Man, according to our theory, prepared in advance and was virtually already contained in his anthropoid precursor when changed environment forced him into the channels of the new human principle—an opportunity that the other descendants of the same parent stock did not have, or in any case missed.

The same consideration should also apply to the very first "great step" in biological evolution, the step from the lifeless to the living. Although about this spectacular event of *pre-Cambrian* times, perhaps a thousand or more million years ago, we know next to nothing, still, for everyone who, in common with the majority of biologists, refuses to believe in a supernatural creation of Life nor would expect Life to have come to the earth by the invasion of mysterious micro-organisms floating through the interstellar space, it is a matter of scientific conviction that Life arose from the lifeless on evolutionary lines, that is, by the operation of the ordinary laws of Nature; and here again we must assume that the great step had its material basis prepared in advance, and was ready to be taken at the critical moment of fortuitous external circumstances.

Modern research work into the construction and composition of the living cell has been most successful in discovering the ultimate particles upon which Life is based—highly complex amino acid and nucleic acid molecules. As they appear to be the actual carriers of Life we are led to believe that from a primordial aggregate of such molecules Life started upon its dynamic development. Gradually it spread over the whole globe in an infinite number of varying forms, all still in close interrelation, interdependence, and interaction, thus forming a coherent category of its own and differing essentially from the non-living world. With Life, then, something new must have come into the world, a new mechanism or scheme due to a new principle in order to transform the

---

the case if mutation was exclusively a chance-event primarily independent of biological need and function. However, the probability of the Whale having by chance developed its aquatic equipment whilst still living on the land and then, conveniently, taken to the seas is too remote to be considered seriously.

aggregate of chemical molecules into a biological *conglomerate*, a self-regulated integrated unit, in short, an *"organism"*. The conception of "organism" as an entity fundamental to the understanding of evolution has been ably set out by A. N. Whitehead, who, however, stretched this concept so wide as to include in it also the inanimate things, such as atoms, which he consequently endowed with a rudimentary form of mind.

We do not know how the first living substance or organism came into existence, and we do not know the new principle which was responsible for its creation. We only know, or believe we know, that the secret of Life lies hidden in the chemical structure and combination of the gigantic molecular structure of nucleotides and proteins. There has been much speculation as to the beginnings of Life. One suggestion is that the first living thing, perhaps through its special molecular arrangement with new properties, had become able to absorb small particles of similar proteins, and on the one hand to integrate part of them into its own structure, and on the other hand to give off, as "waste", the non-assimilated particles, and that by such feeding upon other matter it was bound to expand, and with increase in size and bulk to become capable of division. Emergent Life would thus have carried in it the germs of self-maintenance, metabolism, and reproduction.

However it was that the new principle acted upon its preexistential molecular basis to transform lifeless matter into living substance, it must in any case have been a *natural* event, that is, proceeding along evolutionary lines. Hence, we should expect Life-manifestations to obey the same basic laws that we find operating in the rest of Nature. There is indeed nothing in the physiological processes of, say, breathing, metabolism, bloodcirculation, reproduction, that would not be amenable to explanation in terms of physico-chemical properties, and it is in the face of such mechanical interpretation that many biologists have come to believe that Life is but a matter of mere physics and chemistry.

So we read in Sir Charles Sherrington's work *Man on his Nature*, 1940, a brilliant exposition of the mechanistic points of view: "Broadly taken there is in 'living' nothing fundamentally other than is going forward in all the various grades of energy-systems which we know." And furthermore: "All of that impressive growth and development which produces a child from an egg-cell is in the main chemical building, operated, regulated, co-ordinated and unified chemically."

It may have been from their strong opposition to the "vitalistic" school and its invocation of a non-material principle as being responsible for the creation of Life that modern biologists lay stress on those truly physico-chemical processes which serve the maintenance of Life, such as oxidation and metabolism, whilst they disregard, more or less, those Life-manifestations which are essential for Life and clearly missing in lifeless matter, such as heredity, regeneration, adaptability.

H. Driesch and other "Neo-Vitalists", on the other side, have shown that at an early embryonic stage of development each component cell had its special function in the formation of a definite structure or organ; but if one of those cells had been destroyed, then another cell, originally designed for another function, would step in for the destroyed cell, so that to all intents and purposes the "wholeness" of the growing organism would be re-established. This observation led Driesch to the conclusion that the living cell was endowed with an autonomous behaviour, a kind of "prospective potency" which was no longer open to interpretation on purely mechanistic lines.

Driesch's argument seems to me to run in a circle. It starts with the phenomenon of integrated wholeness as an essential feature of the living organism and ends with the same phenomenon in its manifestation of regeneration as no longer referring to a harmoniously correlated and co-ordinated whole but now representing a novel and inexplicable thing.

His real problem was, as it is still ours, that of *Life* itself. The modern purely mechanistic conception of Life has, to my mind, created a confused situation similar to that created by the conception of Man on purely zoological lines. Here, as there, it would seem, the existence of fundamental similarities, which in themselves can only attest to common *origin*, has wrongly been taken as evidence of common *nature*, too. Yet, inanimate matter is not any more identical with living matter than is the Ape with Man —lifeless matter just lacking what is distinctive of Life, and the Ape just lacking what is distinctive of Man.

From the standpoint of the present book, there seems to be a distinct analogy between the principle underlying the evolution of Life and that underlying human evolution. Both are taken here to be natural evolutionary principles, and as such started upon their evolutionary career from a certain pre-existential basis, the one from inorganic matter and the other from the Animal, and gradually built upon their old basis a new realm of its own category, the realm of Life and that of Man respectively. It is from this condition of emerging from, and still dependent on, the old basis that their evolutionary products, Life and Man, are still explicable, even at advanced evolutionary stages, in terms relating to their old basis, that is to say, Life in terms of physics and chemistry, and Man in terms of animal anatomy and physiology. More, however, than such general reference to the old basis we should not attribute to those terms, and any reasoning from them on the true essence of Life, or of Man, must expose us to fallacies. If we want to describe Life, or Man, as to what they essentially are, we can do so only in terms of the distinctive manifestations of their underlying evolutionary principles. So are the phenomenon of regeneration, such as the healing of a wound, or the growth of a lost limb in lower Animals, or the phenomenon of reproduction, such as the development of a child from an egg-cell, although they are still amenable to description in terms of physics and chemistry with regard to the physiological mechanism involved, yet no longer explicable in such terms with regard to their essentially being innate, ordered, self-regulating Life-processes as manifestations of a new principle.

*

The upward movement of Life, leading persistently and consistently from the primordial living organism, over the "great

H

steps" of animal evolution, to higher biological levels of integration and finally to Man himself has been interpreted in the light of a *directional trend* making, on the one hand, for greater control and independence of environment and, on the other hand, for fuller growth of mind. This is certainly true of Man who enjoys a considerable measure of independence of environment and is endowed with extraordinary mental abilities. It is not so, however, with the Animal, which even in its higher forms remains largely dependent on its environment and has developed its mind only to a degree corresponding to the rise of its biological level. To all appearance, animal mind merely serves and is completely consumed in the immediate requirements of life at the various biological levels. To the Animal then it makes, in principle, no difference whether it stands on the lower level of, say, the Amoeba or Mollusc, or on the higher level of, say, the Bird or Monkey, since at either stage of an integrated system it is equally well equipped, physically and mentally, for its particular pattern of life conduct in its special environment.

With Man it is different. Only through the miraculous ascent of the biological levels up to the Primates was the material basis reached as the essential prerequisite for his emergence and development. If it is right to say that Man not only forms the apex of the pyramid of living beings but is the very crown of the creation, it follows that evolution in its upward movement virtually *worked towards the rise of Man*. Some authors, like J. Huxley (*loc. cit.*), have taken pains to show that natural selection would by itself solve the problem of the peculiar *directional trend* leading upward to Man in that the selectionist process would sustain any adaptational trend running in the direction of evolutionary progress. Even if this be so, the directional upward trend, so profoundly important for Man and practically so unimportant for the Animal, cannot yet adequately be accounted for on such mechanistic lines which refer only to the machinery involved in the upward movement without taking into account the deep significance of this upward movement for the existence or non-existence of Man.

The question thus remains still open: Is Man but the chance-product of a mechanistic process of which he is just the uppermost stage, or is he, after all, the predestined goal, the promise and fulfilment of the mysterious process of evolution?

One may urge as an objection that this is a problem not to be solved by science, as it clearly transcends the scope of scientific method. This is true enough, and only confirms our point that biological methodology might not grasp all the evidence available. In its momentous task of investigating the laws operating in Nature, and resolving things and phenomena into their ultimate elements and basic concepts, science has given a brilliant and coherent account of the structure of the Universe, and assured us that Life is built upon lifeless matter, and Man upon the Animal, and that Life in general and Man in particular are the natural outcome of a process of evolution explicable on mechanistic lines. More, however, than the analysis of the measurable and its expression in quantitative terms science cannot and will not give, and we are left wondering how to account for the other side of Nature, which eludes scientific method—the non-measurable ordered quality of things in their significance by themselves and for the whole.

"Evolution", for instance, is the basic concept upon which science has successfully built up its biological system, yet of which nothing can so far be said as to what this concept essentially means, signifies, and implies. Even the very fact of existence is still the same mystery to us as it has been in ancient times. We may now be able to understand the mechanism by which a child develops from an egg-cell, but are still utterly ignorant of how it is that from the chromosomes, complex chemical molecules, the human body in all its details emerges.

"What a wonder", said the father (in one of Dostoievsky's novels) when having his first look at his new-born child. "There is no wonder in it," replied the midwife, "it's all but development." So what for the one, looking forward, is all wonder is for the other, looking backward, simple mechanics. Scientists would insist that the analytic insight into the nature of things and

phenomena with the aid of basic concepts must satisfy our quest
for the knowledge of our world, since we cannot possibly obtain
any more knowledge by truly scientific methods.

Their rejection of attending to matters not amenable to
scientific exploration has persuaded a growing number of intel-
lectuals into a complacent agnostic attitude. Such agnosticism,
however, has been attacked for its unphilosophical and rather
perfunctory outlook on Nature with no satisfactory answer to the
problem. If scientific *analysis* does not yet reveal the whole story
of Nature, we are reminded that it be complemented by philosophical *synthesis*. The scientist, therefore, should join hands with
the philosopher who, unprejudiced by mechanistic thinking and
not so much concerned with the materialistic aspects and
measurable contents of the world, is rather inclined to look upon
things from a spiritual point of view, and to conceive Nature as
an ordered and indivisible whole with a significance of its own
and ultimate issues in its train. Evolution he would take as de-
scribing a scientifically established fact, and at the same time as a
means of allowing Life to climb up its organic and mental ladder
to its highest rung, Man. He would reflect upon the observational
fact that evolution reveals a directional trend towards the development of mind and, along with it, towards the creation of
Man, and meditate upon the evolutionary implication that Life
has with Man become self-conscious and has lifted him upon an
ethical plane on which the pursuit of truth, goodness, and beauty
seems to transgress the ordinary ends of individual existence.

In this connection he would pose before him the question as to
whether the ethical conduct of Man, his free personality, his
awareness of being an integrated part of the universal evolutionary process, and his belief in his own perfectibility may point
to a higher destination of Man entailing new responsibilities and
the moral obligation to carry on evolution by striving for his own
individual perfection and the perfection of the human society as
a whole. He would thus take up those questions of supreme concern for Man which science is reluctant, or incapable, of approaching; and in his conviction that a purely mechanistic inter-

EVOLUTION AND METAPHYSICS

pretation of Life would not yet resolve the problem of human existence and of human conduct of life, he would endeavour to re-establish, by synthesis and evaluation, the ideal wholeness of the Universe and the ideal harmony of Man with his Universe. Whatever be his ultimate interpretation of the "meaning" and the "significance" of Life, and of Man as representing its highest form, the "soul" of the earth, he would insist that Man is truly Man and in perfect harmony with his Universe only when he is truly ethical in his conduct of life.

"All religious effort", we read in A. Campbell Garnett's book *The Mind in Action*, "is essentially the same—the effort of the finite life to live in harmony with the infinite reality in which it is immersed." Therefore, "when Man feels that his life is in harmony with his universe he can go his way with confidence".

*

The scientifically established fact that there has been a universal process of evolution on the earth embracing all life including Man, and explicable on purely mechanistic lines with no necessity of invoking supernatural forces, has led to the idea that, in the face of our new knowledge, it was about time to dispense with the superstitious and outdated beliefs of former times and to contemplate a "World without God". At the close of the last century it was *Monism,* based on the phylogenetic unity of Life, that was propagated by E. Haeckel as a substitute for the traditional religious systems. Nowadays, *Humanism* is proclaimed as the scientifically ascertained form of religion, a kind of "scientific theology". Here, as there, Evolution is professed to take over the role of God, and mechanistic thinking is entrusted with the task of deciding the ultimate issues.

Considering the deep mysteries that, in spite of all our knowledge regarding the structure and the ways and workings of Nature, are still overhanging the phenomena of evolution, of the growth of Life and Mind, and of the reality and significance of human existence, one may wonder whether any such proposition of gross materialistic tendency, with no metaphysical cloak to cover up the bare bones of scientific analysis, would be apt to create a spiritual *vacuum* with undesirable consequences.

The scientist may be content with his task of reducing the Universe to a system of differential equations, and the song of the Nightingale to a series of sound waves of various lengths. The ordinary Man, however, although fully appreciating the analytic work done by scientists, would not be prepared to miss the beauty of the Nightingale's song, or to regard himself as but "a fortuitous concourse of atoms". The gap, therefore, between the facts of science and the facts of experience has to be bridged some way or other, and the world in which we live has to be re-established in its wholeness and its holiness. It is at everybody's discretion to decide which way to go. However, whether he believes in Man as being the evolutionary result of pure chance, or the product of mechanical causality, or the goal of predestination, he has to fit himself harmoniously to this world of his, and to act in consciousness of his privilege of being the crown of life on earth. Even the most enthusiastic atheist should live up to the *fiction "as if" there was an ordering principle* in the world, and Man the promise and the fulfilment of the evolutionary process; and such ethical attitude towards Life must imply veneration of the Sublime and of the Unknown.

If understood in this sense of an ethical guiding principle, Humanism may well prove a useful scheme to be worked out into a valuable philosophical system based on modern scientific thought with no preconceived prejudices.

Many modern philosophers, it is true, are no longer attracted by the great traditional issues regarding Man, Society, and Nature that were so close to the heart of the old metaphysicians and gave them their strength and their distinction. Instead, they are engaged in a kind of analytical philosophy in which ordinary language is subjected to a scrutinizing analysis of its denotation, implications, and ambiguities in order to expose the flaws, misleadings or false analogies. As ordinary thinking, indeed all we know of the outer world, is expressed in language, the choice and the use of words in their various contexts must be of decisive import on the question of the accuracy or fallacy of any statement or proposition, and it is in this respect of clearing language from ambiguities and pitfalls that modern analytical philosophy has rendered a most stimulating and salutary service to philosophical thought.

Of the traditional issues fallen a victim to this modern analytical trend the most prominent were the "transcendental" notions and reflections cherished so highly by the old metaphysicians—such abstractions as those referred to the *absolute*, the *ultimate reality*, and so on. Although it should be clear that issues of evaluation and of looking at things from *above* rather than from *below* are by their very nature inherently dependent upon inferences by analogy and deduction, those rather vague and unverifiable generalities could not stand up to the scrutination of analytical inquiry.

This must not, however, be made, as has sometimes been done, a convenient pretext for abandoning metaphysics altogether. For if metaphysics were a forlorn post nowadays, yet modern philosophers are still ready to shoulder their responsibilities in giving an account of the world as a *whole* and guidance in the conduct of human affairs, they are bound to face the great problems of the world which are in fact the very things that our inquiring mind is eager to know about—the more so as scientists, in spite of all the progress they have made in their inquiries into the substance of the world, are becoming increasingly conscious of the limits and limitations to their investigations and explanations. Philosophers are thus forced into a position in which they are asked to take up their work of synthesis where scientists have left off, and in doing so and with all the new scientific knowledge of the world on their hands and in their minds, may soon find themselves in the deep waters of metaphysics where they risk drowning if they are not prepared to swim.

Only in the perspective of the process of evolution can Man be judged as to his physical make-up and mental endowment, and to his relation to the animal world and, for that matter, to the earth itself. In this perspective he appears neither small since he has worked himself up to the top of creation, nor a saint, since he is still firmly tied to earth and afflicted with the animal inheritance of vegetative functions and low emotions. Still his greatness lies in the evolutionary event of his divorce from the Animal, whereby he committed himself to a course of evolution in which his animal inheritance was greatly superseded by a new and superior style of life, making him a new and superior creature on earth.

We have some good idea of how all this happened, and so we are sure that it actually *did* happen. But *why* it happened we do not know and for this reason we are the more intent upon the philosophical evaluation of the evolutionary creation of Man, its meaning and its significance, so as to give us hope, faith and guidance.

Man's capability of looking at things in a detached way, and leading him to divorce his sense-organs, his mind, and even his own ego in isolation from, and in opposition to, the rest of the world has brought on some intriguing dilemmas which in the perspective of this abstract detachment seem insoluble, but may resolve themselves if looked at from an evolutionary angle.

One problem is that of the "reality" of the world. It stems from the inescapable stricture that all we know of the world we know through our sense-organs. This subjective element in our knowledge has consequently led to the proposition of setting up an "idealistic philosophy" culminating in a flat denial of the reality of the world—in a sense that there was no tree, no sun, no lightning and thunder, if there were no one who could perceive them. As Berkeley has put it, "To be is to be perceived."

In opposition, there are the advocates of a "realistic philosophy" who on the self-evidence of the senses believe in the "real existence" of the world, yet in their empirical world-view are equally subjected to the restriction that all their knowledge of the world is ultimately derived from subjective sense-perception.

How does the problem show in the light of evolution?

Whatever particular role mutation and natural selection may have played in the development of new organs, it was most certainly in connection with, and in response to, new biological needs that new organs developed---the finlike organs of the Whale in response to the biological need of swimming in water, the legs of the Amphibian in response to the biological need of walking on land, the sense-organs in response to the biological need of obtaining vital information about the surrounding world.

Although, under the present condition of fully developed sense-organs, it may seem that it was our eyes, ears and noses which primarily sensed the world, as if it was our own perceptions that gave the world its existence, it was, conversely, in the perspective of evolution that the world by stimulating the development of sense-organs implanted its image into our brain and which therefore existed before there were sense-organs developed, and will continue so to exist whether or not there are sense-organs to perceive it. It follows that the phenomenalism of idealistic philosophy is wrong, and that the "naïve realism of common sense" is right.

Perception is by its nature direct and immediate, hence it can only affect things in the present, not in the past or future. On behalf of the opposite view of an indirect and deduced character of perception, the objection has been raised that in the case of a star a million light-years away we would see an object that was not actually in the place where we were seeing it, and which might not be even any longer in existence. The objection is invalid because the light through which the star makes itself perceived had to travel all that long distance before reaching our eyes and because of the very immediacy of perception shows the star still in the position from where it originally came, whether or not the star had meanwhile changed its position, or even had ceased to exist.

It poses, however, the question: to what extent do we actually perceive the material world through our sense-organs? In so far as this question can be answered by way of analogy, we may reasonably assume that our sense-organs are biologically as adequately adapted to the necessary knowledge of our surroundings as are the fins of the Fish adapted to swimming and the wings of the Bird to flying. Still, our perceptual picture of the world,

although what it reveals is the "real" world and should be sufficient for our ordinary needs, may not yet be accurate and complete in itself, and indeed proves to be most imperfect if checked
by scientific methods. But however greatly science has increased
our knowledge with its devices of telescope, microscope, spectro-
scope, etc., and introduced us into its esoteric world of atoms,
molecules, gravitational and electromagnetic fields, scientific
vision of the world can never supplant "natural" perception. A
"tree" of whose intrinsic texture we may now know a good deal
is still for us the tree as we naturally perceive it, and which we
use for building chairs and tables; a "tiger" which we may know
down to his every chromosome is still for us the tiger as we see,
hear, and smell him, and which we would have to shoot if he
attacked us. A "star" of which we may know something concern-
ing its movements and fate is still for us the star as we now see it
in the sky, and which we may use as a navigational aid. In other
words, we continue to perceive things by those palpable qualities
that once gave rise to new biological needs and to the subsequent
corresponding development of sense-organs. Thus we still perceive water as the liquid substance that we drink lest we die of
thirst and in which we swim to avoid drowning—rather than as
a chemical compound of hydrogen and oxygen, molecules and
atoms.

Another problem, long since hotly discussed and still highly
controversial, is that of the relation of the Mind to the body. The
Mind–body problem seems to me again to arise from Man's
capability of looking at things in the world in a detached way,
thus taking Mind in isolation from, and even in opposition to, the
body. Yet it is evident that the Mind is lodged in, and harnessed
to, the body, and is influenced by bodily states as much as the
body is influenced by mental states. Moreover, observation and
experiment have shown the anatomical "seat" of the Mind to be
in the brain. The narrower question therefore is, whether Mind is
what we would call a "function" of the brain, or whether according to the old doctrine it is an entity of its own, essentially hetero-
geneous to the body and only loosely attached to, and working

through, the brain. What makes the problem even more difficult is the uncertainty of what Mind exactly is, since in Man it is commonly hidden behind a high wall of inscrutable thoughts, acts, and feelings.

In trying to define Mind in a simple way, we exclude first all those functions and qualities which are obviously none of its constituents, that is, the automatic reflexes and any kind of sensations such as the feeling of pain, hunger, cold, giddiness, and also any kind of volition and emotion which, although more or less prompted and determined by states of Mind, are effects rather than ingredients of it. There remains then only one quality: thinking, and Mind is thus defined here as *the sum total of thinking*.

This definition must include the Animals as well, since they, too, are known to possess the faculty of thinking, and here it is where *evolution* comes in.

In the single-celled Amoeba there is not yet a true Mind–body problem, although its manner of discerning and capturing its food and of reacting to light, chemicals, contact and obstacles may suggest some kind of awareness, however faint. Nor is there such a problem at the evolutionary stage of the Jellyfish where differentiation of structure and function has led to the development of sensory cells stretching out to form a collaborating nerve-net. The problem first arises with the Flatworms in which we meet with a central nervous system, a two-lobed brain, to which the two eyes are connected by extended sensory cells, or nerves.

In principle, there is no difference in kind whether, in the case of the Protozoa, awareness occurs in the sensitive protoplasm itself or, with the Coelenterates, in special sensory cells joined together in a nerve-net, since in either case sensitivity and awareness are still one; or whether with the higher organizational level of the Flatworms awareness occurs in a centralized nervous organ, the brain, which still remains in close connection and co-operation with its sensory cells, such as the light-sensitive cells now equally concentrated in special organs, the two eyes.

That is to say, perception and awareness are fundamentally the same thing. As this rudimentary form of primitive awareness signifies the first beginnings of "psychic" activity, no polarity can exist between perception and Mind, however greatly Mind will increase its complexity with the higher Animals, and still more with the case of Man in the process of evolution.

Neither can any polarity be construed between organ and function. The philosophical axiom that Mind belongs to another category than the body, because the body is "extended in space" and Mind is not, disregards their integral biological unity. Function is essentially a Life-manifestation, an *organ in action*, as it were, and as such existing in, and with, the organ, and not somewhere else. To say that the eyes are "extended" and seeing is not, or that the legs are extended in space and walking is not, reveals stark unbiological thinking. Indeed, organ and function, eyes and seeing, legs and walking, brain and thinking are fundamentally identical in the same sense as Life and living organism are identical.

The Mind–body antithesis thus breaks down under the weight of biological argument. Its collapse does not, however, necessarily imply the total invalidity of the old dogma of the heterogeneity and immortality of the human "soul". Apparently, the concept of "soul" is in contrast to the animal Mind, with special reference to the *rational* thinking of Man and the dominating influence of Reason upon his conduct of Life. Thus, with Reason, the exclusive human principle comes into play, and not only draws the line against the Animal, but in its extra-bodily essentiality allows the thoughts, works, and deeds of Man to survive even after his death —altogether a biological set of ulterior consequences that must invite meditation on the transcendental distinction of the human soul.

A third problem is that of the relation between the individual Man and Mankind as a whole. It embraces the wide field of anthropology and sociology, and can here only be touched upon from a biological point of view.

Evidently, the species exists and maintains itself through the

individual, and the individual depends in his existence and maintenance upon the species which presses him to make any effort to
secure his life and to propagate. This is, however, not a simple
reciprocal relationship. The species is the whole, the individual is
part of the whole; the species works itself out in the individual and
will continue to exist, the individual works directly for himself
and indirectly for the species, and will perish. Modern psychologists have rightly emphasized the prevalence of the species when
they speak of the "id", or the "collective consciousness", working
in Man. It is the "genius" of Mankind that reveals and per-
petuates itself in the works of our great thinkers and artists,
although they themselves did their work on their own and are
justly praised for it.

The mysterious interplay between individual and species con-
forms with the wholeness of Life and, in a wider sense, with the
wholeness of Nature. It again reminds us, in spite of the mechanistic aspects of the Universe, not to lose sight of its philosophical aspects. That is to say, we must look at Nature, in the
words of Spinoza, "sub species aeternitatis", in the perspective of
eternity.

Index